User-centric Design

USER-CENTRIC DESIGN

DIANE
STOTTLEMYER, Ph.D

Westphalia Press
An Imprint of the Policy Studies Organization
Washington, DC
2023

Westphalia Press
An imprint of Policy Studies Organization
1367 Connecticut Avenue NW
Washington, D.C. 20036
info@ipsonet.org

ISBN: 978-1-63723-942-1

Cover and interior design by Jeffrey Barnes
jbarnesbook.design

Daniel Gutierrez-Sandoval, Executive Director
PSO and Westphalia Press

Updated material and comments on this edition
can be found at the Westphalia Press website:
www.westphaliapress.org

CONTENTS

FUNDAMENTALS OF USER-CENTRIC DESIGN, CONCEPTS, AND PRINCIPLES

What is user-centric design?

User-centric design is a process where users work with the developer, and ultimately, the successful application meets the user's needs. Designers and developers use various methods to gather the information, including user interviews, collecting data from surveys, or even setting up a focus group to discuss user needs. Also, user center design is not only for the traditional computer but also for tablets, Smartphones, and mobile apps. Fundamentals that are a part of setting up a user design include usability, accessibility, reliability, desirability, speed, stability, and security.

Usability is an essential component of design that involves users and utilizes their input to create two data types. First, the users' ability to interact and provide feedback increases the designers' awareness, improving the design's value. When working on the design, feedback can reflect how the user communicates their evaluation of the system or application. User feedback and their critique of the scenario through interaction with the design will help make usability improvements look easy. In addition, usability testing can help evaluate the exchange and provide feedback to the design and development team. So by using the interaction between data and feedback, improvements can be used to manage a functional design (Folstad, 2017).

Accessible Design

The design also needs to include accessibility for everyone; for example, any physical or cognitive disabilities need to be developed based on accessibility designs using WCAG 2.0 (Web Content Accessibility Guidelines) guidelines. In addition, the web has made a significant impact on accessible web design. Because of this, a user-centric approach is beneficial to balance the design with the website's usability (Byerley & Chambers, 2002).

Designers discuss accessible design and how users must be aware of accessibility guidelines. Since users are part of the testing process, they are also responsible for ensuring accessible applications that are useable and user-centric. "Educators are becoming more aware and vocal about accessible design issues that concern the disability community" (Byerley & Chambers, 2002, p. 169). This awareness has helped develop the accessible design and supports accessible websites for all users.

Furthermore, in addition to accessibility, as with any application managing data and information for individuals with responsibilities on a day-to-day basis, the application needs to be reliable. Users want an application that can provide reliable data and information and functionality, reliability, and validity for the information gathered and collected with the application. Therefore, the application needs speed, reliability, and stability to capture and process data quickly.

Last but not least, the application design needs security measures that keep the information protected and ready to use when needed. For example, suppose the design developed has individuals' names and personal information. In that case, information needs to be protected to provide privacy to those who may obtain and use the information. Again, users can give feedback on this type of information and the level of security required.

As a user designer, developing a process helps focus on specific tasks that must be completed and adapted to the organization's business. For example, determine the individuals using the designed application and question them on what they need it for and how often they will need it. Then, the users must provide specific requirements that meet and achieve work objectives. Requirements are based on what the users want in their application design.

Designers can include the user in different development phases to ensure they capture the requirements set forth by the users and agreed upon by the developers and designers. In addition, the user needs to be part of the design testing process and determine if revisions are needed based on requirements. This process will help reinforce the usability and user-centric evaluation of the software (Usability.gov, 2021).

Critical Thinking:

Is there a difference between Usability and Accessibility or design planning and development?

Examples of design processes

Depending on the selected design process, the user is crucial in their role. Therefore, there should be a discovery phase that accounts for clear goals and meets users' needs as part of the process. For example, the user would add input to most phases in an SDLC (Software Development Life Cycle) monitored by all parties. Following are different lifecycles which are presented that show essential elements of development with the user playing a vital role in the lifecycle:

Software Designs

Software Development Life Cycle

The basic software development process includes seven steps that help designers and users go through a systematic approach to design. One of the essential parts of software design is planning. Brainstorming and focus groups are beneficial tools for this process as they meet the user's specific needs and requirements. After the planning phase, review information and requirements to ensure that all requirements have been considered.

Then the design is discussed, evaluated, and set up for the development phase. User-centric design is needed to make the application functional, useable, and user-friendly. The developer uses this information to ensure the application is ready to develop before coding begins.

Users will inform the developer of errors and improvements by scripting the steps to work through the interfaces during the testing phase. After implementing the application, the user and designer may want another test to ensure it works correctly. The information is stored and used to improve or modify the design. The maintenance and continual improvement process will be necessary as time goes on for improvements and correcting areas that may need refining. Following are steps for design which include the user.

- Plan and brainstorm – Include the User
- Analysis of the plan and setup requirements – User included
- Design – Include the User to make sure the design is based on the requirements
- Develop and code

- Testing – User Acceptance Testing

- Implementation

- Maintenance and continuous improvement – Ask Users if this meets their needs

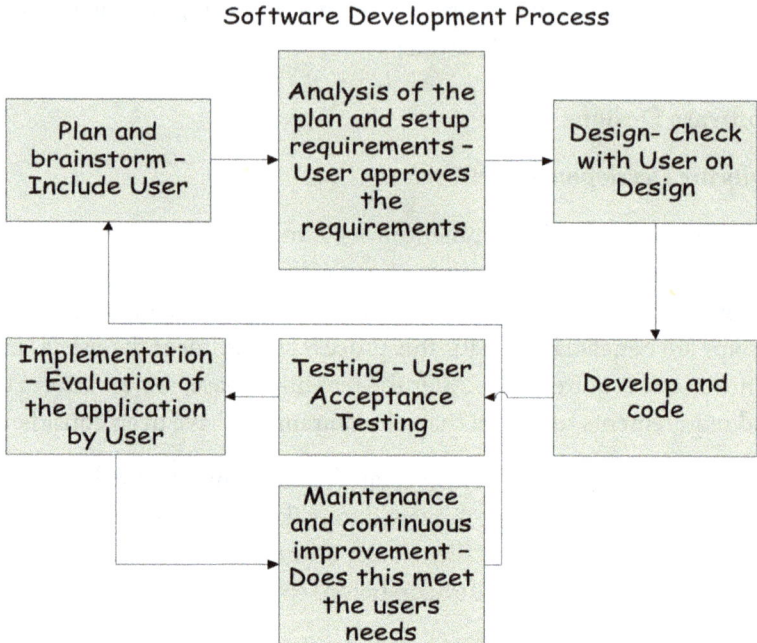

Software Development Process

Figure 1. Software Development Process

Waterfall Model

When working with the waterfall design model, steps are successive before moving to the next phase. For example, the requirements phase is completed before the design phase. Thus, the user would work on the requirements and review them with stakeholders before presenting them to the developers for the design process.

Once the requirements are reviewed and accepted, the user would provide the developer with the requirements and then discuss them to ensure a clear understanding of the user's vision for the application. Once the development process begins, requirements are considered complete unless an error is detected when developing and running a prototype. User acceptance testing will help identify any gaps in design before deployment. Once deployment occurs, timelines are set up to include maintenance and changes that need to be made to the design.

- Requirements – Work with the User on developing requirements
- Design – Check with the User to make sure the design is what they need
- Development
- Testing – User acceptance testing
- Deployment
- Maintenance – Additional changes – including User

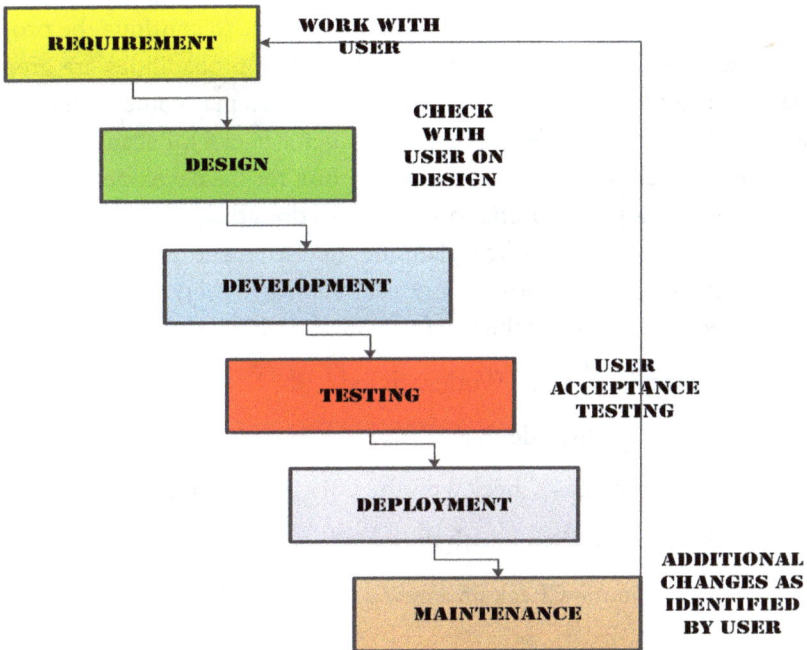

Figure 2. Waterfall Method

Agile Software Development – The Agile software development process is quite different from essential software development; however, there are many similarities in how development works. One of the first differences is that the customer/user is key to the entire development process and success with Agile development.

Often with the many different development lifecycles, the user has a role, but with Agile, they are also engaged in the development process. For example, with the project initiation, the user provides what they want by giving the developer and stakeholders their requirements and discussing how they want the application to work and perform. Often the project begins, and requirements provide information on continuing the project. However, on the website by Cleverism (2021), it is noted, "There is no assurance that all the necessary information or requirements will be complete and on hand at the beginning of the product development process" (para 8).

The project could begin with a handful of requirements to get the project going, and then adjustments may need to be made to continue the project. Also, Cleverism (2021) discussed how "the probabilities are great that, in the middle of the process, new information may come in that will have an impact on the design of the product" (para 8). Because of this, user involvement continues to be a focal point for the developer through every step. So as project initiation begins, so does the participation from the user and goes through the planning, development, production, and retirement phases. This process is a user-centric process for application design because the user is the key focal point of the design.

- Project initiation – Work with the User
- Planning – Include User
- Development – Check with the User as developing
- Production – Work with the User
- Retirement – Check on a new design – check with the User

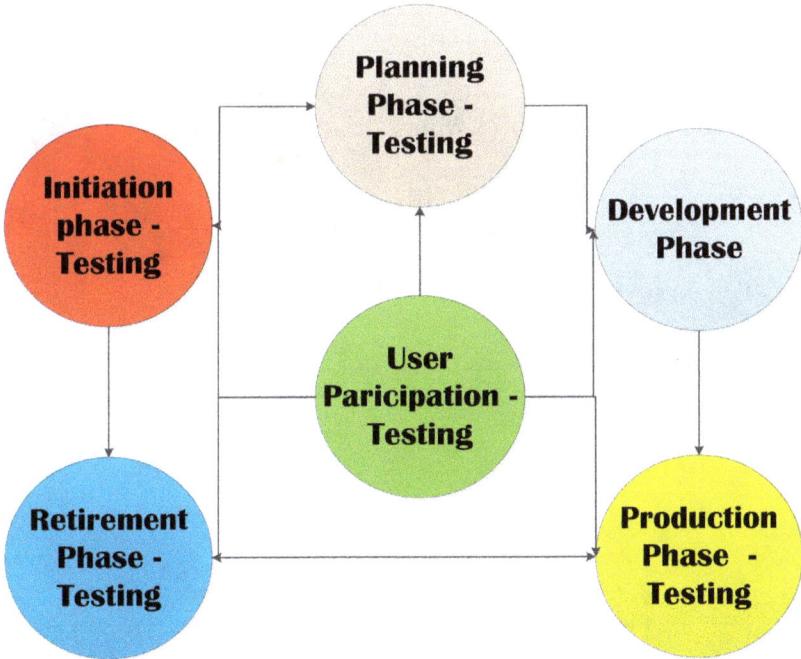

Figure 3. Agile Lifecycle

Spiral Model

The spiral diagram is for projects managed by evaluating the design and is handled in phases to ensure that the design and development are progressing in a cyclic approach. First, prototypes are developed as the design progresses to test functionality. Then, phases are planned based on releases related to specific functionality developed and tested before moving forward (TechTarget, 2021).

Phases are essential for large projects and help minimize the risk of starting over. The phases include a design phase, managed by understanding the requirements and working through a risk analysis before development begins. Identifying the objectives for the application under development is necessary before progressing. The risk analysis helps to evaluate possible problems before they occur. The developers then develop, test, and plan for the next design iteration. This cyclic development process continues until the development is complete (TechTarget, 2021).

- Design – Work with Users

- Code – Developers
- Test – Acceptance Testing – Users
- Implement – Recheck with Users

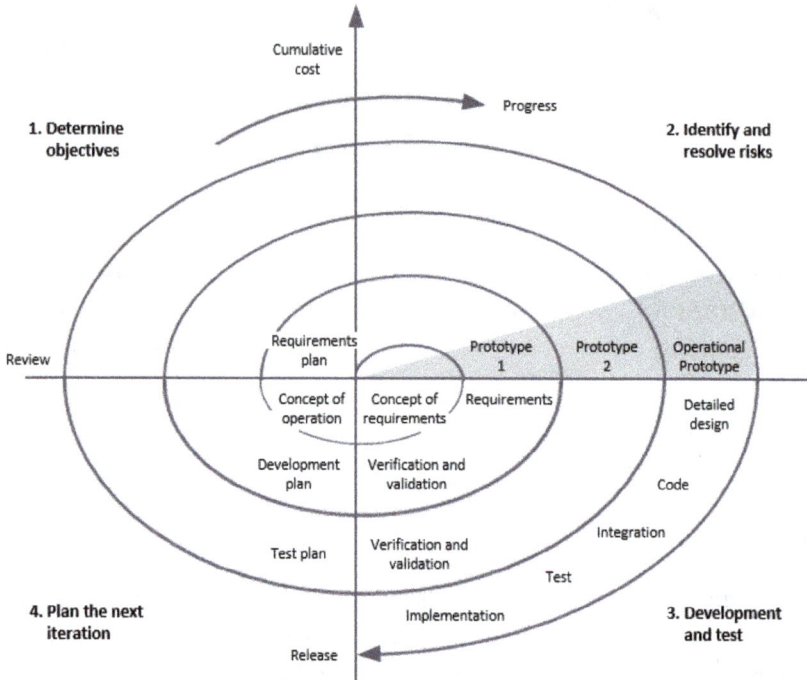

Figure 4. Spiral Diagram
https://www.praxisframework.org/en/library/development-life-cycles

V Model

If you look at the V software development life cycle, you will see that the phases incorporate testing in the different phases of development. This model shows that the user can become essential to the development process. Much like the spiral lifecycle phase, this development phase requires verification and validation before progressing with the development of the application.

The requirements phase is one of the first phases, and also note that the system requirements are needed to plan system reactions to certain conditions. The design is based on the requirements and the system requirements to develop the module design. When these requirement phases occur, coding begins with the developer, and testing is necessary to ensure all requirements occur. The user and developer work together to conduct the testing. After all testing occurs; the final step is to sign off on the application and implement the application. Thus, the left side of the V is the Software Development Life Cycle, and the right is the Sofware Test Life Cycle. In this model, the user plays a crucial role in testing, making this a user-centric approach to development and testing (Guru99, 2021).

- Left – Specification/Requirements – Work with the User to develop the requirements

- System requirements/high-level design – Check on the system to complete the system

- Architecture Design/detail design – Evaluate the architecture – User verified

- Module design/production – Look for consistent design from module to module

- Coding – developer

- Unit test – Developer

- Right side – Acceptance test – User

- System test – Developer

- Integration test – User

- The final step is Implementation/acceptance – User

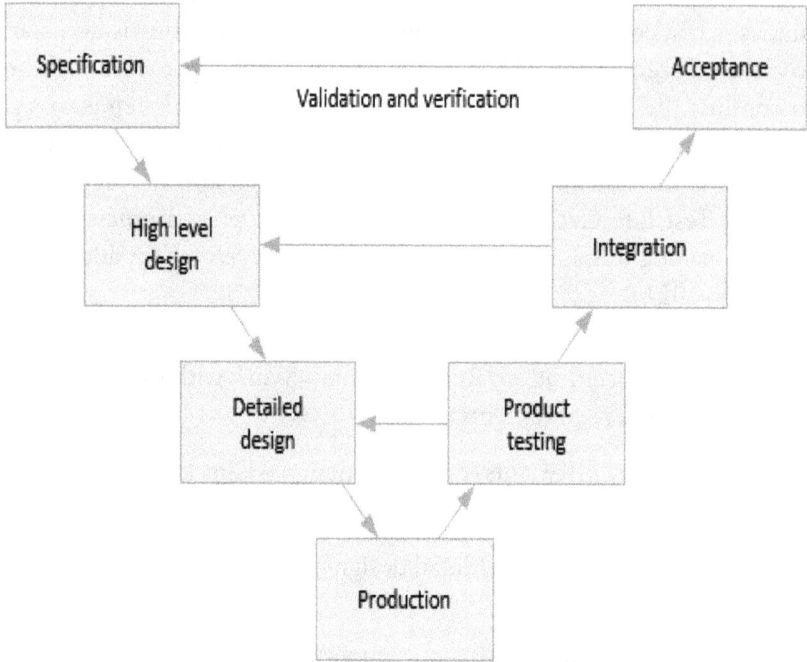

Figure 5. *V-model*
Reference: https://www.praxisframework.org/en/
library/development-life-cycles

Usability Design

Another practical design that includes users is the user-centric design, noted next as a usability guide developed by usability.gov.

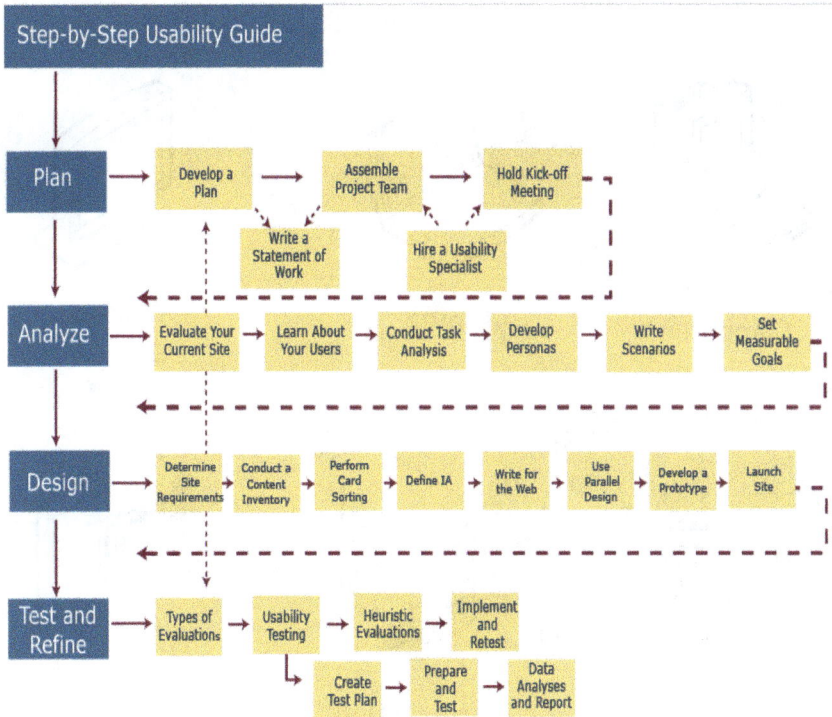

Figure 6. Usability Guide
Usability.gov (2021). User-centered design basics. Retrieved from htps://
www.usability.gov/what-and-why/user-centered-design.html

As noted in this diagram, there is a plan, analysis, design, and testing. First, the planning phase includes the user and shows a usability specialist. Next is a session to learn about the user's analysis phase before writing the scenarios and setting goals. Finally, a prototype is developed during the design phase before testing begins. Finally, a usability test with the user evaluates the results and determines if a retesting occurs, or if the implementation is ready (Usability.gov, 2021). This approach to development includes the user throughout the phases as an essential part of the development process to plan, design, evaluate and test before implementation.

Critical Thinking:

Are models important during the design and development phase?

Personal Computer

Laptop

Monitor

Workstations

Mobile Unit

Tablet

Figure 7. Different Operating Devices

Whether developing for a personal computer, laptop, PC Monitor, several workstations, or mobile and tablet devices, looking at different operational components is needed. Each machine operates differently, and the development process will differ. When developing an application, the cloud environment in today's technology will become part of planning, design, and operation, mainly from storage and a service approach. The user needs to understand how cloud technology may impact active availability and application use. The cloud environment can be used and made to include laptops, monitors, workstations, or mobile and tablet environments. The key is understanding the development process; for example, setting up a mobile environment versus a tablet environment would be different, as reflected in the coding process.

The user-centric design utilizes an analysis that includes who, what, when, where, why, and how.

Jang, Ko, and Woo (2005) suggest the following:

- a particular user – (Who)
- is in a specific location (Where)
- in a specific time (When)
- paying attention to a specific object/service (What)
- making a specific expression with physical signs (How)
- A particular intention or emotion (Why)

Who, what, when, where, why, and how are questions to ask when analyzing the development of an application. User-centric evaluation based on these questions can be helpful in the progress of working through different phases. For example, before beginning an application development, analyze who will be involved and who will benefit. Then where will the application or product be used? Will this occur during a particular time frame, and what specific service or object will be affected? Additionally, how provides the physical application, and why will be a good question to analyze how to solve a problem. Each part of the analysis helps in focusing on a specific need or area where a problem may exist and be solved.

As you can see, working through a process using a specific model unique to different situations can help set up a framework for beginning a user-centric work relationship between the users and the developers of a project. Chapter Two will present information on requirement development and its relation to user-centric design.

Reflections – Chapter 1

Agile versus traditional development

Customers want immediate, quickly developed applications and programs with continual technology improvements and changes in our fast-moving society. That is why the Agile development process has become the norm more than the exception. Therefore, evaluating how Agile differs from traditional development and analyzing whether specific areas lack traditional techniques is essential.

Agile development has six phases: concept, inception, iteration, release, maintenance, and retirement, compared to a traditional waterfall method with requirements, design, development, testing, deployment, and maintenance. So you can see differences in some areas, and the user's role will also change. The project will generate a better, fully developed, user-centric approach if the user can contribute to each phase of the software development lifecycle.

Compare and contrast the two styles and utilize the best approach for the user.

Traditional Lifecycle	Agile Lifecycle
Requirement - Testing	Concept - Testing
Design - Testing	Inception - Testing
Development	Iteration - Testing
Testing	Release - Testing
Deployment - Testing	Maintenance - Testing
Maintenance	Retirement

Questions to consider:

1. Do you want the application quickly or focused on a traditional lifecycle process?

2. Is the application based on user needs or organizational needs?

DEVELOPING REQUIREMENTS – FUNCTIONAL/ NON-FUNCTION – BASED ON USER NEEDS

Requirements – Why are they important?

Requirements provide the designer with what the user needs from an application. Therefore, developing requirements is often referred to as requirements engineering. Requirements engineering (RE) should be a part of the development of any software project. This is important because requirement analysis will feed into the different stages of software development (Jia & Capretz, 2018).

However, additional requirements also lead to scope creep, and the project is not controllable when continuous changes come from the user. So when there are additional requirements, there needs to be an agreement on whether the requirements are needed. In addition, extra changes lead to a perception gap in how the requirements were initially developed and managed. Jia and Capretz (2018) suggested that there may be gaps in what users and designers perceive as requirements, which may cause a barrier in development. The gap can inhibit the development and placement of the requirements into development and may cause disagreement on the direction of the application.

Suppose the user is not confident in what is provided by the developer. In that case, they lose confidence in the development. This gap does not encourage development, and the user loses confidence and may not respond proactively to critiques in the development phases.

Often developers are concerned about the requirements and whether users understand how the development may inhibit the project's success and even the development scope (Jia & Capretz, 2018). Therein lies problems with the development when there is a disconnect between the user and the developer.

As you can see, working with the user helps to define, narrow, and control the development. No one wants to work through the development phase only to have an application that users cannot use. Therefore, designers and developers should gather requirements based on the users' needs that create a user-centric design. An essential part of the requirements missed is the signing off before development occurs. The user and the developer need to agree on the requirements so that when development begins, there is a clear definition of the user needs, and scope creep does not become an issue.

Developers create applications with the user in mind, so a working relationship is necessary between the user and the developer. Therefore, the developer needs to interpret the users' requirements into a consistent developmental format based on the design, which may mean discussing the intention of each requirement. An example of user interaction with developers is with Agile programming. With Agile programming and development, the user plays a crucial role in the process and helps to ensure that what is asked for is pivotal to the application's success.

Generally, requirements are specific standards for the user managed by the developer and are either functional or non-functional requirements. The requirements guide the development process and the documentation manual on operating the system.

There may not be a formal document for requirement submission by users for developers as some requirement-gathering methods may include interviews, questionnaires, social analysis, prototyping, and requirement reuse, as noted by Dar, Lali, Ashraf, Ramzan, Amjad, and Shahzad (2018). However, to ensure a cohesive approach to requirements gathering, there should be a document that can be used to record and list the requirements, so there is a helpful format for everyone on the teams. The user can participate in the interview and give information about the application and what they need to receive. The questionnaire will provide information based on their knowledge of the system. Also, as part of the process for eliciting requirements, the evaluation of social analysis may include the user's political and social participation based on social scenarios. The key to this approach is to communicate and collaborate and then coordinate the evaluation of user needs and verification of user needs versus developers' requirement elicitation (Dar et al., 2018).

Functional Requirements

Software requirements are considered either functional or non-functional. For example, an accounting application's functional requirement may be to ensure the application can keep track of clients and their invoice data. A non-functional requirement may be maintaining the security of the data. The key to understanding a functional requirement is to ask what you want the system to do. Then, in turn, the system will utilize the data entered into the system and then provide the output based on a process or even a calculation as part of the coding by developers (QRA, 2021).

Functional requirements are based on what the user wants from the application. For example, when users want to use Word, they open Word and have a blank screen to type their information for a particular task. In addition, if the user wants to add information to the screen, modify or even delete it, they will type it using the keyboard or mouse. These are all considered functions that are requirements the user wants from the system.

The non-functional requirements are system performance. For example, when the Word application is selected, the user wants to know if it will open and how quickly the system will respond to their input and data. User requirements versus expectations is an excellent way to look at functional vs. non-functional requirements (ReQtest, 2021).

The response time is based on the selection of the keypress and the system's response rate to the keypress. In addition, models may be used to analyze non-functional requirements to help determine the system's specific timing, which is examined further in Chapter 3.

Critical Thinking:
Are users only responsible for functional requirements?

Non-functional Requirements

Non-functional requirements are based on the system's usability, performance impacts, and user expectations (QRA, 2021).

How are requirements developed?

Gathering requirements occur using different techniques. For example, interviews that elicit information from users can be a step that can include focus groups or even group-thinking sessions. Also, it may be helpful to send questionnaires to different departments that know what they want from the developed application. An essential component of requirements gathering may also come in face-to-face interaction between the other users and the developers. The key is that those involved need to understand the needs and the best fit for the business and how the new application can provide the information.

Additionally, the social context in which the application will be used will also impact how the application will be received and managed. Therefore, the administration and primary stakeholders must agree on a plan and design before development (Dar, Lali, Ashraf, Ramzan, Ajad & Shahzad, 2018).

Vener (2019) suggested that "user-centric—business users should be enabled (emancipated) to explore, surface and articulate innate, yet elusive, user-centric business requirements aimed at organizational improvement" (p. 482). So critical emphasis is placed on the users as a focal point and organizational advancement.

Involvement in the development of requirements

Dar et al. (2018) noted that one of the most critical aspects of requirement development is to include the stakeholders. Stakeholders include the "identification of the business owner, candidate stakeholder, evaluation and selection of stakeholders, understanding the role, responsibility, and relationships among stakeholders, stakeholder representative, prioritization, stakeholder management strategy, and plan are also important and relatable" (Dar et al., p. 63861). Based on the stakeholder's input, the requirements are what they want to see for the application based on the design. In addition, the requirements set a prioritization strategy because the developer often says you cannot have everything and the kitchen sink. Different challenges can also include "knowledge, scope, change, human factor and organization related" (Dar et al., p. 63861). Developers note the challenges impacting communication between stakeholders, users, and developers. For example, stakeholders may not be directly involved in the application based on budgetary requirements or lack of time to work with the users and developers.

The knowledge-based is then centered around the user and communicated to all stakeholders. Users help by developing and providing their opinions and experience as the software is developed (Jia & Capretz,

2018). Because of this perception, users may lack initial interest in the development process. Therefore, the scope is evaluated, and the requirements should relate to a user-centric design and meet the organization's needs. However, as Jia and Capretz (2018) noted, there are differences in understanding all requirements.

The users have a specific idea of what they want in an application and do not see the entire picture of planning and design. This gap must be addressed in the planning stage before it reaches development, so it does not continue to grow. This means interaction between the developer and user is critical from the beginning and continues through each development phase.

The Agile software development cycle encourages user participation for a user-centric design, so it is often a focal design and development process. The user participates and is a productive part of the team (Jia & Capretz, 2018, p. 278). As shown in Chapter 1, the Agile process encourages users' participation. The users are critical in providing information about what they need for the developer.

> **Critical Thinking:**
> **Why does everyone need to sign off on requirements?**

Developing requirements in each stage of the design

After reviewing each lifecycle process, the designer and user can evaluate which method meets their needs and which both the user and designer agree upon.

So as the user and designer work with the requirements phase of development, they should consider the following:

- Determine what problems will be solved
- Gather input from the stakeholders and users
- Evaluate the problem with the stakeholders and user
- Determine the direction in which to solve the problem
- Develop the requirements that will help solve the problem

- Look for various solutions to the problem

- Select a solution that meets the business needs

- Work on the layout of the design for the user and designer

- Develop a prototype to see if the problem is solved

Chapter 3 will discuss computational models and how the user is necessary for designing and developing applications based on the different models.

Reflections – Chapter 2

Defining functional and nonfunctional requirements

As a user, understanding what a requirement is and why it is essential may be considered a topic that is not user-friendly since software development is not the focus of most users. However, explaining how functional requirements differ from a developer's non-functional requirements is necessary.

An excellent example of the difference between functional and non-functional requirements is considering who will use the application and how the application will work and what both need. The applicable requirements are the basis of a user, and the developer works extensively with the nonfunctional requirements.

The functional requirements are based on business requirements; in other words, what does the user need for the system to function while the developer needs to ensure security, reliability, and performance? You can see how different the two areas are and how the user and the developer need to understand each other's requirements.

Questions to consider:

1. How does the user determine business requirements?

2. How does the developer determine and prioritize non-functional requirements?

COMPUTATIONAL MODELS FOR USER-CENTRIC DESIGN

GOMS Model

Understanding models and their advantage can help determine cognitive users' needs and will help improve application usage. Keeping the user as the central focus involves understanding what the user needs to work with an application. For example, the cognitive aspects of the user may include learning along with eye, hand, and body movements.

Examples of computational models are GOMS, KLM, Hick-Hyman, and Fitts. The GOMS model helps users and developers understand how to process human information. The initials for GOMS mean goals, operators, methods, and selection rules for processing information. Goals are based on users and what they want from an application. Operators are used to show actions where methods take the goals and break them down into sub-goals. The S in goals is to look at competing ways to determine the best approach.

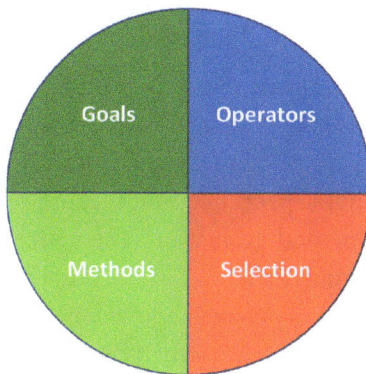

KLM Model

Another model is the KLM model. This model examines the number of keystrokes necessary to complete an action. This model is considered a predictive model for human-computer interaction development. Developers can use this model to predict a user's action and determine how long the action takes to complete. The KLM model includes six operators: keyboard, keypress, point, draw, mental, and system response.

An example of the KLM model may be to examine how long it takes a user to open up a Word document using the menu. Next, developers can use this model to evaluate keypress actions, for example, using the mouse as the pointer on the application. Also, there can be interaction by the user with the keyboard. Further, there may be use or non-use of drawing related to cognitive (mental) learning, and finally, there are system response rates. The developer may clock the different actions based on when the user sits down, turns on the computer, boots the application, selects the program, and begins to use the application. Timing each activity will help the developer understand what type of response rate will impact the overall reaction time and the application's usage.

The developer may need to work beside a user and evaluate each step based on the model. Each step can be timed to determine the impact of the action and how long the activity takes. This evaluation process uses the information to improve the development and speed up or even slow down the activity to accommodate the users' needs. For example, understanding how operators can be used can impact basic cognitive controls, including pressing a key or moving the mouse. This basic cognitive approach of the mouse, keyboard, and keypress is considered the physical aspect, whereas the mental aspect is related to a cognitive evaluation.

An example of a fundamental physical motion is the key press, release keypress, pointing, and changing from the mouse to the keyboard. These are all examples of physical responses in operating the system. When considering the mental aspect, decision-making plays a crucial role in evaluating cognitive mental evaluation. The system response times are based on system operation and how long it takes to occur. Timing of each area would be beneficial to determine how long each function takes and whether a modification is needed, which can impact the application's design (Model-Based Design - Cognitive (user) Models, 2002).

Fitts Law

When looking at Fitts Law, users play a crucial role in evaluating how they use a mouse pointer to assess a target for the distance (Interaction Design Foundation). Fitts' Law interprets time and distance to complete an action. Therefore, the user is needed to provide their approval for the interface related to speed and efficiency.

The psychologist Paul Fitts was responsible for coining this Law. He felt that human movement accounts for measurements to determine an individual's move to a specific point or distance along with its speed to move to the specified distance. The consistency of movement is evaluated, measured, and calculated for errors in movement. This technique has helped evaluate devices from desktop to mobile. Design and testing will vary based on the size of the machine and the ability to interact with keyboards of different sizes.

Additionally, Fitts Law relates to pointing movements, how quickly an individual moves the mouse, and how the user may have continuous movement when working with the mouse (Interaction Design Foundation). Therefore, by applying Fitts' Law, the user's approval role impacts the design based on their movements and interaction with the application.

An example may be that the designer wants to account for the movement necessary for the user as they interact with an interface. Therefore, the design should be kept simple and easy to use. Another example may be the command buttons; they should be basic and easy to read, see and click. The click from the mouse will generate the pointer, and then by releasing the pointer, the designer can calculate the time necessary to complete this action. Users generally have a defined area where they can point and click on an application. Evaluating this interaction and movement can help determine the necessary time to complete the process.

Fitts Law noted that "as the size of an object increases, the selection time goes down, and as the distance between the user's starting point and the object decreases, so too does time" (Interaction Design Foundation, n.d, para 11). Thus, time and design play a key role in calculating the time it takes to set up and work with a design.

Hick-Hymans Law

"Hick's Law model is also utilized in user design. Designers often say, "Keep it Simple, Stupid," to avoid overwhelming potential users or website visitors. Keeping the design simple is a fundamental principle in user-centric design projects. The user needs to have a quick and easy-to-read interface readily available, and they do not want to look around the screen to find the correct button to press.

Interface design and web interfaces must be graphically pleasing and easy to navigate. "This is especially important for websites when planning navigation menus. Having too many links makes it difficult to find the category the user is looking for and discourages them from staying on the website" (Whatis.com, 2021, para 4). Setting up navigation windows can be evaluated to keep the information presented in an accessible format. It would be beneficial for the designer and user to look at several formats of user interfaces to see good and bad designs to improve the design and make it user-centric and useable. When evaluating reaction time, it is necessary to account for age, gender, and stimulus.

Developers have the flexibility to manage human-computer interaction with the user in mind. For example, the developer codes to get interaction from the code, and the user inputs their information to arrive at a specific location. Therefore ultimately, the user accepts and interprets the information in a format that will help them complete an objective and goal for their job.

In Al-Megren, Khabti, and Al-Khalifa's article on *"A systematic review of modifications and validation methods for the extension of the keystroke-Level Model"* or the KLM suggests that this computational approach elicits information from the user to achieve selection rules characterized by GOM or what is considered goals, operators, and methods. Timing of the activity is the key and can be based on the time needed to achieve the required action from the user. Timing is the practical approach to design, as the quicker the action occurs, the faster the results will happen.

Timing also aligns with behavioral and psychological events for the user. They must process the information to understand the length of time to complete the action. This mental processing can achieve the desired results (Al-Megren et al., 2018). This is referred to as using predictive modeling, where data is evaluated and compared to a baseline of activities. Data is assessed on the length of time to complete the activity and whether changes in design or process may help improve the activity. The key instruments for the KLM are the mouse and the keyboard. The user takes the mouse and controls the action, and the motion and timing for this activity are recorded to determine efficiency.

The activity between the motion and timing occurs using the keyboard and the user's ability to select the correct key to see how quickly the information can be accessed and processed. It is important to note that typists who type fast would undoubtedly have a different recorded score than slower ones. When the tests occur, individuals with like speed may be needed to demonstrate and evaluate the samples for a comparative sample (Al-Megren et al., 2018). Also, the conscious ability may play a part in assessing the timing of the test. Those individuals who understand the application may process the information quicker than those who do not understand the process or the application. An example may be to take a specific part of the program to see how long it takes different sample sizes of individuals to process the information, keeping like components of the individuals constant, then take a sample of dissimilar sample sizes and conduct the same test. This may show variables in similarities, differences, and even outliers.

Following is a comparison between the four different types of Computational Models to show the similarities and differences.

Table 1. *Computational Models Compared*

GOMS	KLM	Fitts Law	Hick-Hyman Law
Goals	Keypress	Time	Design
Operators	Point	Mouse control	Reaction Time
Methods	Keyboard	Target	Movement Time
Selection	Draw	Distance	Response Time
	Mental		
	System Response		

GOMS – Goals Operators Methods Selection

KLM – Keypress, point, Keyboard, draw, Mental system response

Fitts Law – Time, Mouse control, Target, Distance

Hick-Hymans Law – Design – Reaction Time, Movement Time, and Response Time

PASSAT framework

Models are essential, and so are frameworks. A framework can provide a view based on valuable information structure for problem-solving. Myers, Tyson, Wolverton, Jarvis, Lee, and desJardins (2002) discussed a user-centric planning framework. The framework is called PASSAT, which means there is a plan authoring system based on sketches, advice, and templates. Each term in PASSAT relates to a user-centric approach that combines tools for a plan for users. Templates are provided for managing and sketching the plan to determine solutions as part of the planning process. Users can utilize the templates to evaluate planning, which helps develop systems or applications. The templates that users select can be useful in assessing hierarchical tasks and operating procedures basic to a specific task and provide different ways to analyze and evaluate tasks and subtasks. Outlines can be created through sketching that helps others provide opinions on the plan and the process. This planning approach includes users, who become more involved by providing their input into planning and sketching tasks. Myers et al. (2002) shared an example where the plan was used, and there were different icons in the folder to illustrate the tasks. The example also provided the user with information requirements for the planning process. By incorporating the template, users can see how tasks are broken down into a format where tasks, subtasks, and constraints are evaluated and measured.

> **Critical Thinking:**
> **What are the critical elements you would add to your model?**

Proposed User-centric Computational Model

As you can see from the different approaches to managing user-centric design, there are many different ways to evaluate how users respond to applications throughout the development phase. However, each of the previous models discussed provides specific elements beneficial to analyzing user behavior.

Following is a suggested approach to evaluate user-centric designs that combine the software lifecycle and computational models. This proposed model approach draws on the importance of software development design and user-centric involvement:

Table 2. *Lifecycles and Computational models*

LifeCycle Model	Computational Models – GOMS, KLM, Fitts Law, Hick Hymans Law
Phase 1	Phase 1
1A. Plan	Develop Goals
1B. Requirements	Operators – show action needed
1C. Analyze	Take goals and break them down into specific methods
1D. Design	Evaluate competing applications
Phase 2	Phase 2
2A. Prototype	When working with the prototype, analyze the time it takes to work with the application – KLM – keypress, point, keyboard, draw, mental, and system response
2B. Test	Evaluate the time (Reaction, Movement, Response), interface, design
2C. Implementation	When implementing the application, continue to evaluate time and movement for efficiency
2D. Maintenance-Continuous Improvement	Review information from users to determine if specific areas need improvement

- Plan – Develop Goals

- Requirements – Operators – show action needed

- Analyze – Take goals and break them down into specific methods

- Design – Evaluate competing applications

- Prototype – When working with the prototype, analyze the time it takes to work with the application – KLM – keypress, point, keyboard, draw, mental, and system response

- Test – Evaluate the time (Reaction, Movement, Response), interface, design

- Implement – When implementing the application, continue to evaluate time and movement for efficiency

- Maintenance – Review information from users to determine if specific areas need improvement

- Continuous Improvement – Continue to assess and manage up-grades and new additions to the application. (See Figure 1)

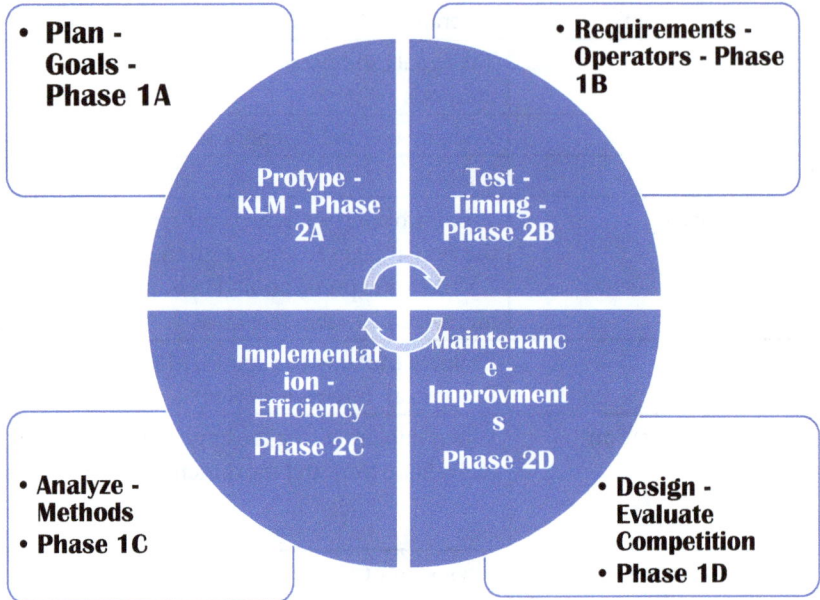

Figure 1. *Proposed Model User-Centric Design Model for Development and Computational Approaches*

This proposed model provides a combination of the computational phases along with the software development life cycle. The alignment of each lifecycle and computational model stresses user-centric design. Furthermore, the management between each phase helps the user understand each development phase the developer goes through to set up the initial application.

In Chapter 4, we will look at the importance of planning for the design and how psychological principles impact the design and development of the application.

Reflections – Chapter 3

Models – Are they beneficial?

In this chapter, we looked at models and evaluated their importance. For example, are models beneficial, and do all models provide the user and the developer with answers to questions on design and development and whether the information is accurate and reliable when following a model's design?

If a comparison is made between the different models, information can be evaluated based on specific goals and objectives the user and developer want to obtain based on development. For example, the KLM model provides helpful information for understanding how a system reacts to user interaction. The Fitts Law focuses on time, target, mouse interaction, and distance to complete an action. The Hick-Hyman Law evaluates the design, movement, reaction, and response time. Each model based on what the user needs can help define a specific goal and objective when assessing the application's or product's overall success.

Questions:

1. Would a combination of each model be more beneficial than one specific model?

2. How would a user develop a process that would mirror each model?

CHAPTER 4

PLANNING FOR DESIGN BASED ON
PSYCHOLOGY PRINCIPLES

Design and Development plan

The design plan helps designers and developers understand user expectations. By using one plan, developers can provide their thoughts and perspectives on the requirements that lead to the design. User inclusion is necessary for the project's beginning, middle, and end. The goals are essential and will help the developer determine whether the requirements obtain the goals and objectives of the product design.

The plan is developed on user-centric needs and the fundamental user-centric psychological principles that will drive the development. As noted in Chapter 3, different models evaluate different needs of the user that may involve cognitive reactions to an application design. Bowler, Koshman, Oh, He, Callery, Bowker, and Cox (2011) noted that "The point of user-centered design is not just to create something that works but rather to create something that works for the intended user, something that is usable" (p. 724). The user is the focal point, and requests for the application should center around the user's needs to complete their work and experience a positive behavioral reaction. "Usable designs created to facilitate information practices have a specific purpose: they should make it easy for people (users) to find, choose, use, and share information" (Bowler et al., p. 724).

A useable plan may be developed using who, what, when, where, why, and how to approach it. This analyzing method is used to uncover various aspects of the design process that may impact the overall design. For example, let's say we have a group of users that would like a website or application to help them track the various demographics for an online research program for higher education. The plan for the developer and user is to identify who the users will be, what the application will be used for, when the application is needed, where the application will be used, why the application is essential, and how the application will solve a problem that may currently exist. Following is how the plan can be set up and evaluated using who, what, when, where, why, and how.

Setup the plan for user-centric design

The developer and user analyze requirements to develop a plan using the who, what, when, where, why, and how. Following is how a focus group or a brainstorming session could facilitate the idea for the development of an online website that focuses on research about retention in higher education:

Example: Online Research Program for Higher Education Retention

1. Who – Anyone interested in learning about higher education research.

2. What – Information based on past research experience and research involvement.

3. When – The development of the site/application-developed scheduling tool agreed upon with all in the group.

4. Where – This discussion is needed and would be beneficial for examining the platform and its location.

5. Why – This question focuses on the need for this development type and moves higher education research forward.

6. How – This question focuses on the users involved and any limitations or conflicts.

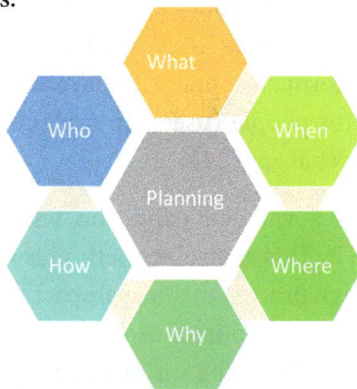

Psychology Principles for User-centric Design
Psychology principles

The development of user-centric psychological principles can guide progress based on each lifecycle step and the method used to collect information from the user. "The methods range from large-scale data gathering, like surveys, to intimate, face-to-face interactions. In addition, UCD (User-Centered Design) is typically iterative in that the design is modified throughout the process so that it increasingly matches the user's requirements" (Bowler and Cox, 2011, p. 726). So this means that the collection of information begins; it may change and be modified based on the collection throughout the process. However, it should not constantly change. Instead, there should be a point where the user and designer agree on the business requirements and develop the application or product.

So, what does the user need to be successful with this application? Several different types of psychological approaches are:

- Include the user in decisions
- Include the user in business requirement development, and let them know they are part of the process
- Does development meet the user's needs? Is it easy to use?
- Keep the design simple and consistent
- Make the interface user friendly
- Make the design specifically for your audience
- Apply accessibility to reach all audiences

Experience (2018) additionally noted that Psychology Principles are necessary to provide users with a design that will meet their expectations and impact their behavior. "You can design your products to elicit specific responses and actions from your users" (Experience, 2018, para 2). When the design is complete, usability is based on the design's effectiveness and how practical the application or product will be. Several factors include:

- Visual perception
- Memory
- Design based on color

These factors impact design. For example, the font may be a visual factor consistent throughout the design. Also, memory plays a crucial role in repetitive use. For example, an individual does not want to constantly search for and use the data. Also, the font color and interface should align with accessibility standards so that all users can use the application

(Experience, 2018). Saghafian, Sitompul, Laumann, Sundnes, & Lindell (2021) noted that ensuring performance and quality match the user's wants. So for the user to have an interactive system and one in which they accomplish their objectives and goals, physical and cognitive approaches are needed to consider it to be user-centric. As users become involved in the design process, they are also involved in the technical approach to making sure requirements are human-centered. Saghafian et al. (2021) further noted that when working with users, they need to look at the overall approach to providing what they need. The example that Saghafian et al. (2001) discussed was when using different technology, such as head-mounted devices; they need to account for motion sickness based on the lack of consistent system timing. This can be an issue and should be accounted for in the design phase.

> ### Critical Thinking:
> ### Should psychological behavior be a part of design and development?

User Experience

Using Psychology User Experience

The user experience varies depending on who is involved and the application developed as part of the development process. "User's approach designs with different intents, varying degrees of engagement, and unique backgrounds of personal experience, beliefs, knowledge, and abilities" (Bowler and Cox, 2011, p. 725). Therefore, users bring experience and on-the-job skills because they know what it takes to do their job better.

Hibbeln, Jenkins, Schneider, Valacich, and Weinmann (2017) noted that "Experiencing negative emotion during system use can adversely influence important user behaviors, including purchasing decisions, technology use, and customer loyalty" (p. 1). For example, if the user is not happy with the design, they will not want to continue, and their loyalty to the project may inhibit the continuance of the application. Also, they may look elsewhere for designers and even companies to work with on the project. An example may be to observe how the users interact with the prototype design using the interface with an action using the mouse. The action may involve seeing how quickly they can interact with the selected area's functionality and if the system's response provides the information or data required to complete the task. If there is a negative response, designers can take this opportunity to improve and make changes before continuing with the design. Hibbeln (2017) noted that when the interaction between the user and the mouse is negative, the user loses their ability to focus on the task at hand, negatively impacting moving forward with the required completion of the task.

Another important part of the psychological impact on user-centric evaluation is eye movement. Jongsoo and Sungwoo (2017) suggested that "gaze analysis using the eye-tracking method to evaluate the usability of major portal sites" (p. 5655). Eye movement can evaluate how a user's eye movement impacts the system's usability. An example of eye movement was evaluated using a monitor and an infrared eye-tracker. The coordinates were measured using X and Y coordinates and evaluated by capturing the movements. This evaluation method offers the designer an awareness of how the movements are tracked and their efficiency. As someone who uses the computer daily, eye movement analysis can be a beneficial addition to analyzing user design since this is a large part of the user-centric evaluation. However, overusing the eyes can lead to a tiring experience working on an interface over an extended period. Therefore, adjustments must be made to make the experience positive (Jongsoo & Sungwoo, 2017).

Looking at good and bad design for users

A bad design may consist of an interface that is not easy to navigate, and when the user tries to find information on the interface, it is hard to find. As a result, the bad design causes the user to leave or not even work with the application because it is difficult to follow. Also, the interface may have content that is difficult to understand and color schemes that make it difficult to view the information.

So, if you are looking at a good design, you would consider the opposite of those mentioned above. Navigation of the interface relies on a consistent design throughout the different interfaces. Also, there is a format to the content that makes finding information accessible, and the color scheme is easy to view and meets a successful design (uxdesignworld, 2020).

Following is an example of what may be considered a bad design (Figure 1) in Wireframes. When you look at the interface, it is unclear on the organization and navigation.

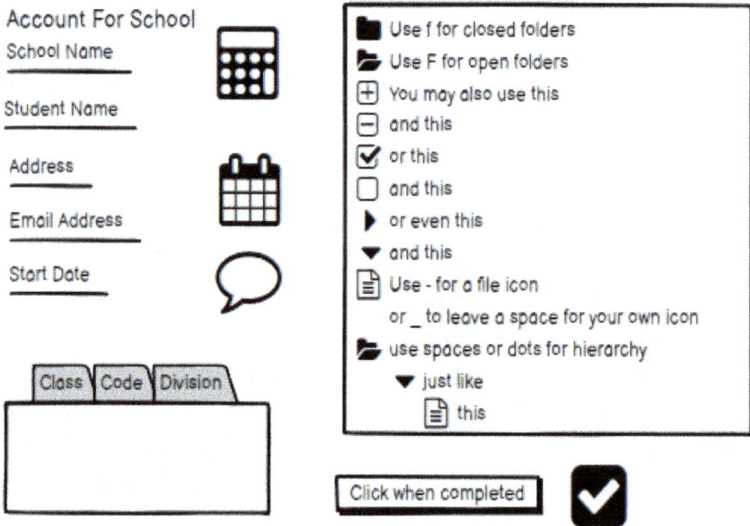

Figure 1. *Bad interface Design*

Following is an example of how designers can change the interface and make navigation clearer. The interface has titles, divisions, and each section to navigate through the interface (Figure 2).

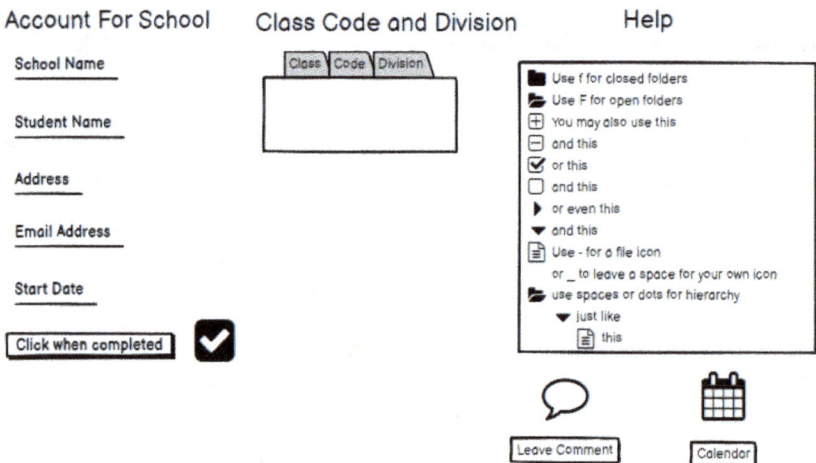

Figure 2. *Good interface Design*

46

Derby (2019) noted that "because of bad design, websites can become cluttered, hard to read, difficult to understand, and frustrating to use. To avoid this, designers often use heuristics, or best practices, to develop an interface" (p. 1). In Figure 1, the interface is cluttered, difficult to navigate, and even difficult to understand the application and the audience. When users come across bad designs, most users throw in the towel and move to different applications or websites. Navigation needs to be consistent throughout the different parts of the application.

> **Critical Thinking:**
> **Are all designs logical for the user?**

Combining Designer, Developer, and User

Merging user design with development

Allowing users to work with developers can increase the impact of success, particularly with the design and navigation of the interface. "Users may engage in interaction and reflection. During the interaction, users' behavior involves the user evaluation of an interactive system or its abstraction, such as a mock-up or prototype" (Følstad, 2017, p.3). As the design progresses, so does the acceptance from the user. This design engagement helps provide an aesthetically pleasing interface and useable interface.

A positive outcome improves acceptability for the user design and the developer. "The user's willingness to adopt a new technical system is based on their expectations of what the system would be like in use" (Kaasinen, Kymalainen, Niemela, Olsson, Kanerva & Ikonen, 2013, p. 6). This suggests that if the user can contribute to the design by providing suggestions and input, they will likely embrace the application's outcome. Also, the user's "expectations arise from experiences and information the user acquires from the world around him/her—what the user perceives in their physical and social environment and what their possible hands-on experiences are with similar or related systems" (Kaasinen et al., 2013, p. 6). If the user has seen other applications and has had positive experiences, they are likely to share this experience and even provide additional critiques on improving design development. Thus, the user will have input that can help enhance and surpass previous application experience, which will help during the testing and implementation of the system.

Kaasinen et al. (2013) suggested that users must be aware of the usefulness, value, and ability to access a user-friendly application. Also, the user needs to feel they are in control and trust the system to provide the information they need and find necessary to complete their tasks. "User acceptance depends not just on technical features but also on the social and cultural context of the user as well as his/her characteristics" (Kaasinen et al., 2013, p. 6). If the user incorporates their particular social and cultural context, this will impact how others will either negatively or positively respond to the application development process.

Chapter 5 will look at the importance of prototype design and provide information on how to set up a prototype and how the user can utilize this.

Reflections – Chapter 4

Is usability important to the user experience?

Psychology principles impact design and development with users—design and development impact how users react to design and development. The user-centric psychological principles can help guide a progressive approach to the lifecycle process. Psychological factors impact how users perceive information visually through memory, design, and color. Usability affects the design, and the users' role is critical.

Questions:

1. Can usability impact decisions on accessibility issues?
2. How will users know when a design is workable even if they agree on the plan?

DESIGNING THE PROTOTYPE – ANIMATION, WIREFRAMES, IOT, CLOUD, AND FOG NETWORKS

Developing a Prototype

"Prototyping is an essential part of product development in companies, and yet it is one of the least explored areas of design practice" (Lauff, Kotys-Schwartz & Rentschler, 2018, p. 061102-1). Developing prototypes helps users evaluate the application's usefulness in a predevelopment mode. Managing the prototype requires knowledge of different tools and methodologies to produce different ways of assessing the application. Developing a prototype and working with a design can help evaluate the following stages in the development process. Design can involve various lifecycles combined with different design tools. "Design work is often considered fundamental to engineering, and thus, observing engineering in practice gives insights to the entire design process" (Henderson, 1991, pp. 448-473). The key is to include all those involved in bringing the application together.

Prototyping allows users and stakeholders to develop a working model to test and utilize. During prototype testing, users can evaluate whether the application functions and addresses the business requirements in the planning stages.

There are different types of prototypes, and one is the rapid prototype. This prototype aligns with agile development and the users' ability to work through each phase with the designer, allowing users to test the application and the requirements. Fritton, Cheverst, Kray, Dix, Funcefield, and Saslis-Lagoudakis (2005) noted that "rapid prototyping is the relatively fast physical fabrication of a design or concept for purposes such as demonstration, evaluation, or testing" (p. 58).

Rapid prototyping is like a jump start to design evaluation and provides a quick corrective action when evaluating the design. Fritton et al. (2005) further noted the importance of working quickly on the prototype; the faster you can determine the weaknesses and make corrective choices, the quicker you can move forward with production. In addition, prototyping allows users and stakeholders to develop a working model to test and utilize before actual development.

It is essential to obtain feedback and evaluate potential new requirements to improve the design (Fritton et al., p. 63). This design and development method helps advance the process and minimize the time for development and possible corrective action or revisions.

> **Critical Thinking:**
> **Should designers utilize sketches from the user?**

Traditional Prototyping and Rapid Prototyping have different prototyping techniques, including paper prototyping, digital Prototyping, and native Prototyping. Paper prototyping is a technique where a pad and pencil are used to sketch the design. A discussion can facilitate corrective action by sketching the design and showing users what they envision. A pen or pencil is used to develop a drawing to show intricate details and corrections by the user and the designer. Digital Prototyping allows the designer and user to create a prototype using various design tools, and the design changes digitally on the screen when modifications are used. The design may be considered interactive because of the ability to access and change the design. Native prototyping is used to manage the prototype design using a real-world approach. This type of prototype is developed and considered a functioning model. It would contain a workable prototype version and evaluate a real-world working model. Development skills for this type of prototyping and testing are done with the user. The user plays a critical role in each prototype model because they know what will work for them and what will not (Babich, 2019).

Design based on user interaction with the designer may often include storyboards, designed sketches, and even models that show the prototype when completed. There are different approaches to developing a prototype. Interaction Design Foundation (n.d) suggests that design thinking is "a non-linear, iterative process that teams use to understand users, challenge assumptions, redefine problems and create innovative solutions to prototype and test" (para 1). This process can lead to a plan to help the designer relate to the user by evaluating the problem and working through a viable solution.

The design thinking process comprises five phases: empathizing, defining, ideating, prototyping, and testing. This process is beneficial because, in our modern society, getting an application that is useable and workable for the required environment may need to happen quickly, so having a design process is essential. The stages noted above do not have to be done in sequence and may occur throughout the design (Interaction Design Foundation, n.d).

Design thinking begins with empathizing with the user. Empathizing means you must understand what the user wants and think of solutions to the problem by designing choices that will fit the user's needs. The second phase is to define what the user needs. Then, isolate the problem and define it precisely so the user can identify with the analysis, creating a tangible problem statement. The third phase is the Idea phase. Creating a workable solution will begin; the user and designer evaluate different viewpoints about the problem and then determine a direction and a solution. The fourth phase is the prototype. In this phase, the analysis and ideas created can help develop a product version, determine if the answer may be beneficial, and help solve the problem. Finally, in the last phase, the prototype can test the problem, where redesign and changes may occur, and alternatives developed (Interaction Design Foundation, n.d). (Figure 1).

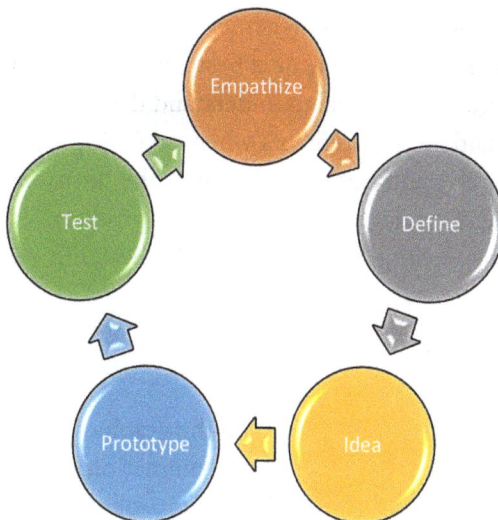

Figure 1. Design Thinking

Lauff, Kotys-Schwartz, and Rentschler (2018) noted that "Prototypes enable communication by creating a similar mental model between people, thus reducing the cognitive burden that can occur during an abstract, verbal conversation" (pp. 061102-5). In addition, by working with the user on the development design, there can be a "meeting of the minds" about how the end application or product should function. If this occurs early, this can help minimize the time spent trying to understand what the user requires.

"A prototype can first create a common language between two or more people. Then, once that common ground is established, used to gather feedback or negotiate aspects of the design" (Lauff, Kotys-Schwartz & Rentschler, 2018, pp. 061102-5). Once the commonality is determined, the design can progress, and there is room for negotiation during the design phase. Design clarification can increase productivity and help minimize revisions and corrections that may have occurred during the design planning process. Lauff, Kotys-Schwartz, and Rentschler (2018) also concluded that "If the prototypes do not work as intended, then they can be deconstructed, or "debugged," to aid in learning about technical aspects. This process can solidify technical knowledge for the designers" (pp. 061102-7).

So once the design occurs, the team of users and designers can evaluate the prototype design and assess whether the design can function as a usable application and help to create a best practice for future designs. So with this in mind, it is essential to understand that the collected information can help validate requirements when preparing a prototype (Lauff, Kotys-Schwartz & Rentschler, 2018). Once the decisions are solidified, they can be used to manage the application or product development. The users can evaluate testing to ensure that the design meets validated requirements.

Animation

"Motion design is the discipline within the graphic design that concerns itself with the question of how to bring animation and visual effects on digital screens" (Rakheja, 2018, p. 1). When looking at the importance of animation with UX design, the key is to look at the design applied to motion. In addition, the motion needs to be accessible and manageable by the user. "Motion design is an important tool in our toolbox for achieving this primary objective" (Rakheja, 2018, p. 1). Therefore, developing a UX toolbox will help evaluate animation for interface design.

For example, "From arrows gliding around a screen pointing you to various screen elements, subtler motion design can help guide a user through an application" (Rakheja, 2018, p. 1). This motion evaluation shows how to manage an interface and how the user can decide how much motion is practical and manageable when navigating the website. In addition, there can be specific automation that will help improve the application's usefulness. For example, "Animation in graphical UI helps improve the user experience by attracting and directing attention, reducing cognitive load, and preventing change blindness" (Pibernik, Dolic, Hrvoje & Kanizaj, 2018, p. 2).

The use of animation helps to work through how individuals operate using the interface and how this will help users understand how the application is used. This cognitive approach involves individuals' interaction using mental knowledge. This is an integral part of user interaction and evaluating animated information.

Wireframes

Wireframes are a newer type of development design tool. This tool has a layout that is like a website page. This format is user-friendly, planned, and designed efficiently using a primary drag-and-drop format. Thus, usability can be evaluated quickly from initial design to development. The Wireframe uses a template format that includes buttons, grids, and text boxes. The format is much like a wireframe; the information created is viewed within a line that looks like wires when designing the prototype layout. Because of the basic blueprint design, designers and planners can look at the design and work with users to discuss the basic functionality. The key to wireframes is to set up a plan and then work through the design to look for different types of required functionality, and no code is used when setting up the basic interface design using wireframes. This design technique can begin with an idea and then be applied to the essential elements of an interface design. Following is a basic setup for design planning using wireframes (Balsamiq, 2021).

For example, Let's say you want to design an interface to collect data for an education group that wants information on recent books for specific classes as an additional resource. Before beginning, discuss who will use the interface, the goals of the site and what needs to be collected. After a basic design setup, a process flow chart can be prepared to determine how the developer will take the design and incorporate the design into the code. So for the initial questions:

1. Who will use the site – Students

2. Goals of the site – Collect books as resources for classes

3. What will be collected – Books to be used as resources

The developer and user can develop more questions and answers based on the complexity of the system design (Balsamiq, 2021). For example, see Figure 2 and Figure 3 below.

When you first open wireframes, there is a blank screen with various icons to set up the design. The image only has some elements available; many more icons and different templates will fit most design plans.

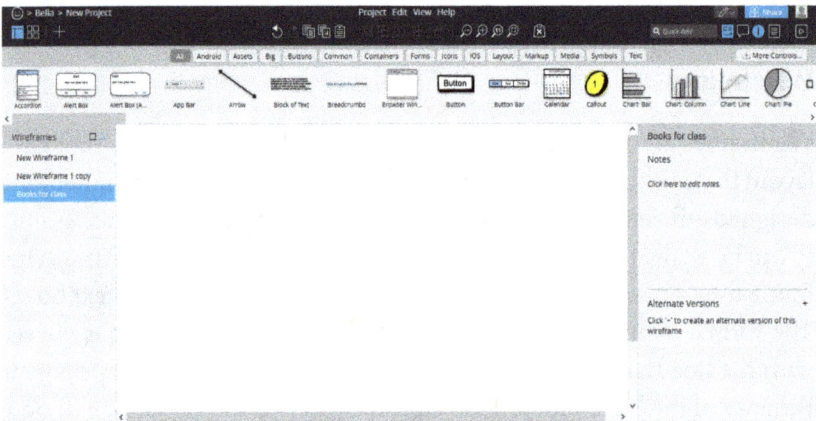

Figure 2. Wireframe Example

The following design shows a basic plan for our example of setting up a project for locating books as a class resource. The basic design uses text boxes, buttons, arrows, and drop-down boxes.

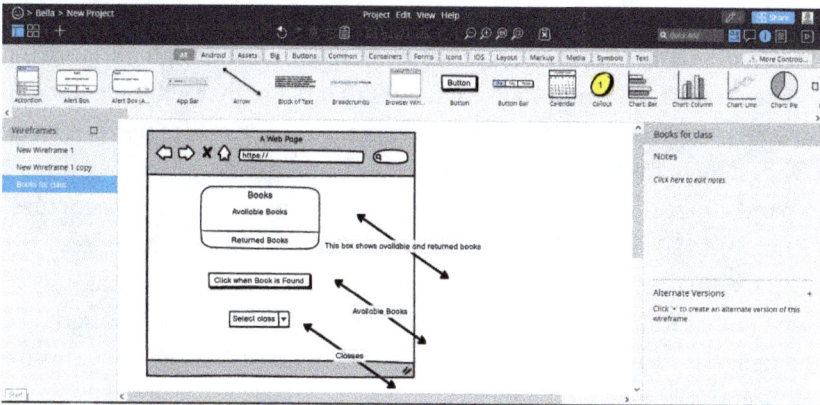

Figure 3. Wireframe Example

Once the design in wireframes has been determined, the mock-up can be designed, and a process flow chart can be developed. There are additional tools that may help in prototyping, such as Figma, Adobe XD, Sketch, and Lunacy. Next, we will look at IOT (Internet of Things) and Cloud Computing.

IoT

The IoT (Internet of Things) is a technology used to manage many projects. The internet of things is an object with sensors; there is a connection and data exchange with devices and systems. Several examples may include lights, appliances, and even cars. The IoT is a communication method using interrelated computer devices with a unique identifier. This method of communication transfers data without communicating with an individual.

The communication process improves communication and makes it faster and more efficient. By using this process of wireless sensors, individuals can attach to an object through the internet to transfer information and data. Not only is this technology suitable for most individuals, but it can also benefit those with disabilities, aging individuals, and user-centric projects. "To harness the such potential, user experience (UX) design must include easy-to-use interfaces, privacy, security, and fairness for all. Without all four, the IoT will experience significant limitations in its growth" (Rochford, 2019, p. 255).

So as a designer, the interface, development of a privacy component, security, and fairness are essential when evaluating the design and looking at how the IoT can help transport users to a plane that will be equitable for everyone. Also, as a designer, "Personalization includes performance. UX designers must design IoT device and app interfaces, so they do not depend on real-time communication with cloud-based infrastructure" (Rochford, 2019, p. 256). This communication process addresses the need for IOT and cloud computing to go hand in hand to manage a design that can follow accessibility protocol and interactivity. Although IOT is a new way to incorporate technology, there are still some concerns for user-centric design, which include:

1. Design may become complex because of the process to connect

2. Technical aspects for the UX Designer

3. Connectivity

4. Complect platforms

5. Connectivity to third parties

6. Trust in using IoT

(Baskar, 2017).

Even though there may be many challenges, this design and development option can still benefit UX Design. "One of the ways to determine excellent IoT platform UX is to simulate the tasks conducted by typical IoT platform users" (Hilton, 2019, p. 1). This simulation of tasks can help users see the IoT platform's advantage and how to design applications' interfaces. Hilton noted that stakeholders could help manage the IoT approach to design by involving their expertise. This may include a platform administrator responsible for the configuration and function of the platform. Then, a platform operator could go through a day using the platform and provide feedback. Also, hardware, software, and backend developers are vital in managing the development, operation, and evaluation of the integration of the system. They are responsible for managing the complexity of the design and development. Then, the user is a crucial member of the process by evaluating the interface and providing feedback and expertise on what could be modified or improved (Hilton, 2019). This process aligns with design and development but includes IoT platform monitoring.

Cloud Computing

Bhaskar, Jukan, Katsaros, and Goeleven (2011) noted that Cloud computing "is a model of service delivery and access where dynamically scalable and virtualized resources are provided as a service over the Internet" (p. 3). It is essential to understand how the service is delivered and how this can impact the implementation of the application. As seen in Figure 4, each application can receive services through the cloud, which hosts servers. The information is stored and accessed when needed. So the goal of the cloud is "to provide on-demand computing services with high reliability, scalability, and availability in distributed environments" (p. 3). In our fast-paced society, this can be a real advantage and ultimately improve operations and meet the demand of users. In a user-centric environment, users must receive information quickly and accurately when they access their devices, and the cloud can be instrumental in this process. "The architectural requirements are classified according to the requirements of cloud providers, the enterprises that use the cloud, and end-users" (Bhaskar, Jukan, Katsaros & Goeleven, 2011, p. 6). The providers have requirements and provide a service based on the software, platform, and infrastructure. The enterprise requirements are based on cloud deployment and whether the service will go through a private, public, hybrid, or community cloud. This service will impact the speed of receiving information. The user requirements are based on how often they use the service, the level of security, service guidelines, and user experience. This service will ultimately have a business process that will impact the quality of service.

So the "ownership of the data is separated from the administration in the cloud. The resources, data, and services are provided to users in the form of virtualization" (Su, Li, F., Shi, G., Geng, K., & Xiong, J., 2016, p. 754). The information collected and stored can be retrieved by the users when needed. Cloud services can be advantageous since cloud storage helps minimize the storage on in-house servers and computers through virtualization. Therefore, the security and privacy of the data are a must, and stored information needs to be secured and managed to ensure privacy. See Figure 4.

Chapter 6 will look at accessibility and behavioral and ethical implications.

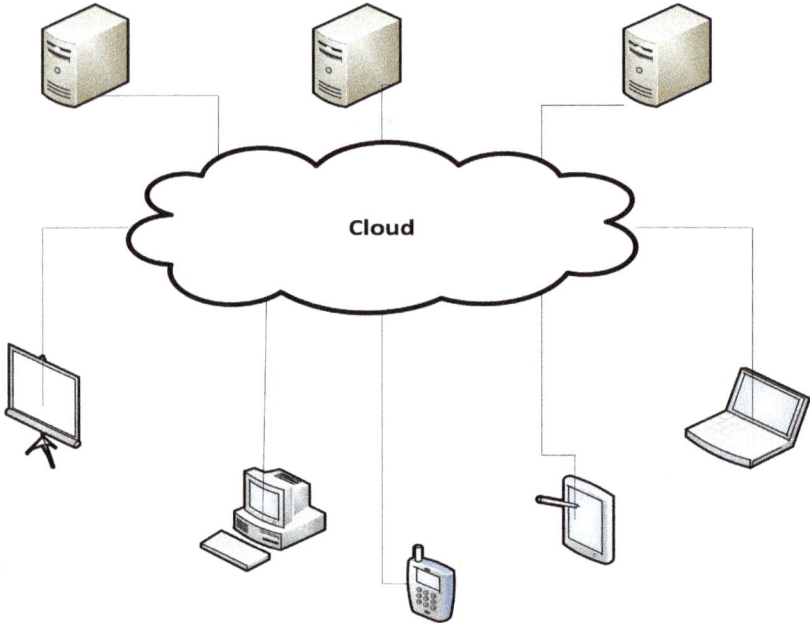

Figure 4. Cloud Services for different applications

Reflections – Chapter 5

Are prototypes essential for users?

Prototypes help evaluate an application before it is implemented as a working application. Before assessing the application, prototypes can be vital in isolating problem areas. Also, the prototype can be tested to determine whether the application can work through the requirements and try the functional requirements.

Questions:

1. Is rapid prototyping better than traditional prototyping?
2. What are the advantages of sketching over rapid prototyping?

ACCESSIBILITY, BEHAVIORAL AND ETHICAL IMPLICATIONS

User-Centric accessibility and behavioral and ethical implications go hand in developing usable applications, particularly for users with a disability. Disabilities may be cognitive, sight, hearing, or physical. The development of applications needs to include accessibility access so that a diversified population can utilize all aspects of the application. Wentz, Dung, and Tressler noted that "there are an estimated one billion people worldwide with a mental, physical, or sensory limitation that could impact their ability to fully use technology (World Health Organization (WHO), 2016, 2017, p. 3). Based on this statistic, a large portion of the population needs assistive help or devices within our society. Accessibility guidelines have been developed, including the W3C (World Wide Web Consortium) and the WCAG (Web Content Accessibility Guidelines).

Examples of accessibility alternatives include additional help accessing information, including alternative ways to display information through text, using captions for multimedia like Youtube, and alternative ways to access information that is not only through the keyboard (Wetz, Dung, & Tressler, 2017). An example of accessibility tools may be screen readers for the blind. A screen reader allows users to listen to information without seeing it on a screen. Examples of screen readers are JAWS for the PC and mobile devices. VoiceOver is a screen reader built into Apple Inc.'s macOS, iOS, tvOS, and watchOS. Both tools are essential for the visually impaired and provide usability for individuals.

Since more and more individuals with disabilities are accessing the Web, usability issues remain essential. "Web usability has become one of the key success factors due to the rapid growth of web application worldwide. It is challenging to create web standards to represent usability practice" (Sohaib, Hussain, & Badini, 2011, p. 586). Setting up a website that includes accessible features benefits user-centric involvement. For example, if the designer works with a user with a vision impairment, the designer can adapt the graphic features to align with the disability. Sohaib et al. (2011) further noted that functionality and content are key areas to look for when evaluating accessibility within the web app. Accessibility equipment can include many different forms. For example, assistive technology that can be useful when using computers may consist of the following:

- Screen readers
- Screen magnification
- Text Readers
- Speech input
- Alternative input, such as
  - Head pointers
  - Motion trackers
  - Eye tracker and single switch entry devices

Table 1 provides a listing of accessibility equipment.

Table 1. *Accessibility Equipment*

Accessibility Equipment	About the Equipment
Screen Readers	Software to read content JAWS – Windows NVDA – Mac
Screen Magnification Software	Enlarge the font and zoom in on different areas
Text readers	Help to read text with a synthesized voice and highlight and emphasize words
Speech Input	Speaking into a microphone while the computer translates the words
Head pointers	Stick or object that is mounted on the head is used to push keys on the keyboard
Motion tracking – eye-tracking	The device that used to target the eyes and place the mouse pointer and move it for the user
Single switch entry device	Useful devices to help with alternative input devices – on-screen keyboards.

Berkeley Web Access (2021). Types of Assistive Technology, Retrieved from ttps://webaccess.berkeley.edu/resources/assistive-technology

Accessibility design

The W3C is a helpful resource that provides information on the importance of User-Centered design and accessibility for all individuals. When working on accessibility issues, the primary approach is to work with users to identify specific usability and accessibility tasks. The key is to look at the design to ensure that all users can access and work with the application. Next, users must feel comfortable using the application and undergo a testing phase using specific prototypes to show usability and accessibility. Usability plays a key role when the application is practical and effective and users like and want to use the application. So basic steps accomplish the usability task, including analyzing the design's goals and objectives before going to development. The next step is establishing a strategy using tools such as wireframes or sketches while using a basic prototype. Then a review of the analysis and design can evaluate the application (W3C Web Accessibility Initiative, 2021).

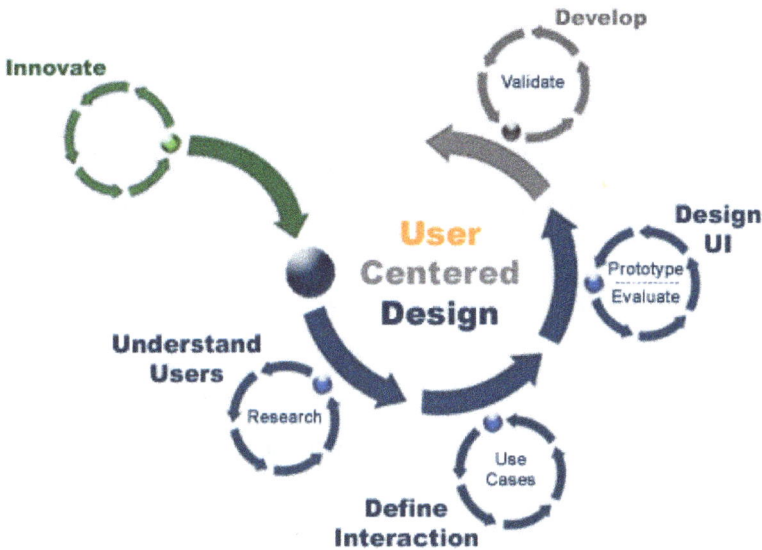

Figure 1. User-Centered Design - User experience design (UXD)
http://insertmedia.office.microsoft.com

As you can see from the design in Figure 1, the user is the focal point of user design. Before the design begins, there is an innovation of thought, objectives, and goals. Next, information is translated by understanding what the users want and researching the best approach. Next, the user can discuss the accessibility issues that may occur based on the design and help the developer by discussing how the issues may impact the overall design. Since the user understands the structure of what is needed and how it will affect users, they can become involved in developing use cases that show each process and the interaction required. Then the prototype design is developed based on the use cases and focuses on the accessibility approach to design for the application before testing with the user.

Another approach – Figure 2

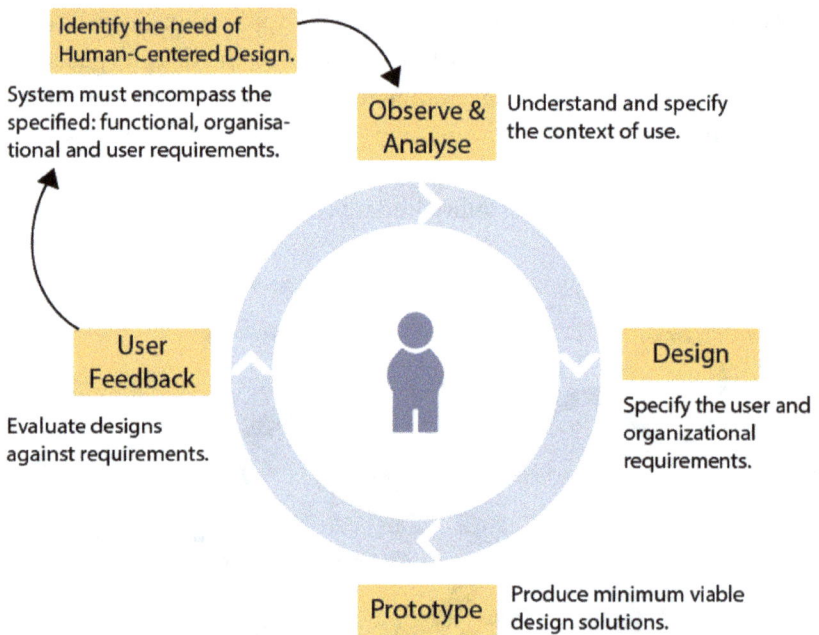

Figure 2. Human-Computer interaction
http://insertmedia.office.microsoft.com

In Figure 2, it is clear that the user is the focus of the application design and the process. Identifying the needs begins the user-centered process, referred to as the Human-Centered design. Designers benefit from this approach by observing and analyzing current design approaches and then looking at improvements needed for future applications. Then, the designer discusses the design and goes over the requirements to make the application functional. A prototype development would be beneficial as the user can see if the plan meets their needs and addresses accessibility issues such as vision, hearing, and cognitive components. The feedback from the user is essential before moving forward with the actual application, which will help improve time and budget concerns.

Behavioral implications

The design and development of user-centric applications involve understanding behavior and usability. Therefore, understanding users' behavior when working with a designer or developer application will help develop an application that meets accessibility standards. In a research paper by Sarker, Colman, Han, Khan, Abushark, & Khaled (2019), the problem of context awareness is based on a predictive modeling approach.

The modeling approach was used to assess diversified behavior and how the behavior was concerned with activities conducted on Smartphones. In addition, a predictive model introduced a data-driven approach to information. The research was built upon a user-centric adaptation that included the Internet of Things (IoT). As discussed earlier, the IoT is beneficial, mainly because of sensors used as a technology for smartphones. It is safe to say that the use of Smartphones has inundated our society and our daily lives, and receiving information by using textual information in a format that is useful for users is based on form and format.

Sarker et al. (2019) offered a behavior decision tree to help predict user awareness of evaluating textual information. The behavior decision tree was completed by working with the device and assessing how individuals interact with contextual information on their Smartphone devices. The decision tree evaluated behavior based on context, behavior, and decision. The model was used to assess temporal, spatial, and social contexts. This model reaffirmed that user-centric context could be evaluated using a predictive model and help determine phone calls and the behavior of a user-centric application. Also, the behavior was assessed based on different situations and diverse behavior that also apply to the age of mobile phones and the internet.

Ren (2020) claimed that more information flows through the internet with the advent of the mobile age. Therefore, user behavior is important to evaluate to determine how access to information will continue to grow along with the users. Ren (2020) further noted that cloud-based mobile usage would continue to impact how data will be used and the behavior associated with the design and architecture related to big data and preprocessing of information.

User behavior can impact how data is processed and communicated. Users want information quickly and accurately and do not want to wait when they have several projects going simultaneously. The cloud has driven one significant aspect of increasing speed and accuracy. When the cloud is optimal and functioning, getting information quickly can speed up integrating information when needed into a design, context, or formative aspect. Ren (2019) discussed an evaluation of the position of the user, location, and social networking experience determines how users interact and "proposed a "cloud-based mobile Internet big data user behavior analysis engine solution" (p.1). Working with big data can be challenging since there are many ways to account for the data and its storage. So keeping the design cohesive with the goals and objectives will enable users to access big data information using cloud technology quickly.

When looking at behavior, conditional factors can be considered. Coyle (2021) noted that different techniques could help understand and influence user behavior. The suggested methods are classical, operant, and shaping conditioning. Understanding each conditioning technique will help when working with users. An example of classical conditioning is if a user hears a bell or ring sound when an application opens up, they will be conditioned to know that the application is opening up when they hear the sound. The user understands that action is required to initiate the response.

An example of operant conditioning would be when the user is working on an application and needs to save. They would select save and then provide the system with the required save command and where they would place the document. An example of shaping conditioning is based on how the behavior is reinforced. So, for instance, after the user understands that clicking the save button will save the document, they know they would have to complete this action to save.

Critical Thinking:

What are the benefits of users input to accessibility?

Ethical implications

Aufderheide (2013) noted that there are ethical challenges when evaluating a User-Centric environment. Trust and faith can be analyzed with three elements: the subject, audience, and sponsors. Challenges continue as a new way of looking at information through digital media. Aufderheide (2013) mentioned that this could occur with different media, such as information on the web, interactive formats, and information distributed on various platforms. The point is whether the information is valid, which poses an ethical question of whether users should be inundated with the information in many different formats. When developing information, integrity and honesty are relevant, and there is a responsibility for those working with the media. Those working with the press must check it out and examine the information to account for its validity.

Urquhart's (2018) article on the ethical dimensions of user-centric regulation noted that two different ethical approaches exist concerning user-centric regulation. The ethical approaches include the involvement of IT designers as it applies to ethical practices and whose responsibility it is to the user. Therefore, looking at the impact of IT design based on user experience is necessary. In addition, when looking at how regulation is carried out, understanding who is regulating the activity will help if additional questions arise and who can help answer them. For example, developers and users ask if specific standards must be adhered to to ensure the information is conveyed accurately and correctly.

In Chapter 7, we will look at the importance of testing and quality checks

Reflections Chapter 6

Who is involved in ethical behavior?

Ethical behavior is essential to understand and implement in an organization. Therefore, standards are necessary and will impact integrity and honesty in decision-making. In addition, the standards help evaluate the

direction according to the guidelines and teach employees the importance of ethical behavior to moral and conditioning behavior.

Questions:

1. Do you think most organizations train employees' in ethical be-
    havior when developing and designing applications?

2. What are the drawbacks to the use of behavior decision trees?

USABILITY TESTING AND QUALITY CHECKS

Testing

There are many types of software testing: unit testing, integration testing, functional testing, acceptance testing, usability testing, manual testing, and automated testing. The user's role is crucial in each type of testing, particularly for acceptance and usability testing. First, the user tests the application, which is considered acceptance testing; through this test process, a determination is made as to whether it meets the user's needs. Next, the user specifies whether they can use the application and complete the requirements for usability testing. Finally, the user would agree that the system satisfies the criteria in user acceptance testing and is usable, and can be deployed. Before the actual test, some organizations may have a beta test to ensure that the conditions mimic real-world applications. The beta test manages the operation of the application before it is put into production and often occurs in the final stages of development. When working through a beta test, the user checks the requirements against the design and development to ensure that the application meets the usability test requirements.

An application's usability improves with a user-centric approach to planning, designing, developing, and implementing the application. In other words, if the user is pleased with the application and it meets their needs, they will be more likely to incorporate it into their daily jobs, increasing productivity. Therefore, a vital component of the development and use of the application is user/usability testing. As you can see in Figure 1, the testing process goes through many different phases. First is the planning process with the user and the developer, when the user goes over the different scenarios and tasks that would make up testing scripts. Next, users will work with the developer through the testing process and use data to generate results and analyze them to determine if the testing meets the requirements. Finally, the tests will establish whether the users will use the application.

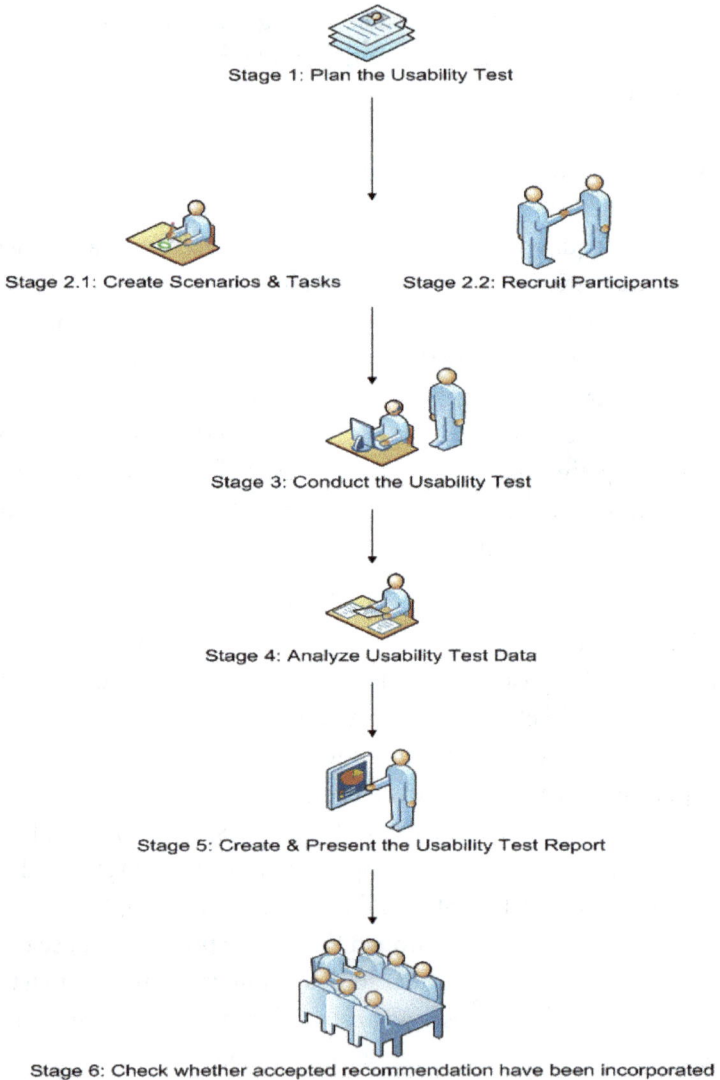

Critical Thinking:

Should users be the primary testers or should there be a team of testers?

The Usability Testing Process
by Abhay Rautela, ConeTrees.com

Stage 1: Plan the Usability Test

Stage 2.1: Create Scenarios & Tasks   Stage 2.2: Recruit Participants

Stage 3: Conduct the Usability Test

Stage 4: Analyze Usability Test Data

Stage 5: Create & Present the Usability Test Report

Stage 6: Check whether accepted recommendation have been incorporated

Figure 1. Usability Testing - http://insertmedia.office.microsoft.com

Figure 2 provides a flow of how a test scenario is developed and how the test scenario flows through different phases. Different approaches can be designed to set up the scenario, but Figure 1 provides a flow that will help the user go through the application methodically. Step 1 is to have a Test ID and provide a title for the scenario. The second step is to describe the test. Then identify the function that you will be testing. The test will help the user identify specific areas for each step and how the user interacts with the application. As the user progresses through the steps, the user will log the expected and actual results. A test script is developed from the scenario that goes step by step through the process. For example, if a user is testing the functionality of a check box on the interface, there are specific steps that the user will need to go through before marking the check box. After the test is conducted, the designer, developer, and user can evaluate the results.

Another example is showing how testing occurs when the toolbar is activated. Figure 2 shows how a test script can be set up to ensure the test process goes through each step. First, the user will need to understand the description and the tool that the user activates and provide a Test ID. Then the flow of the steps would be next. A step-by-step process for starting the toolbar is next. The key is to go through a step-by-step process to ensure that each step is activated. Then the test script will be used to identify issues such as the length of time to open the application, did the measures provide the correct results. The final column evaluates whether the user understood the test steps and whether the results match the requirements.

Test ID - Title of the script	Description	Test Function	Steps for the Test	Expected Results	Actual Results

Figure 2. Test Script

69

The usability testing process can be improved when matching users who use the application regularly. For example, a web application has different testing phases than an application, and other users may or may not use the interface daily. Stumpf (2018) suggested six different usability tests beneficial for different types of tests.

Software testing may include unit testing, integration testing, functional testing, acceptance testing, usability testing, and manual and automated testing. The user's role is important in different types of testing, particularly for acceptance and usability testing. First, the user tests the application for acceptance testing and determines whether it meets their needs. Next, the user indicates whether they can use the application and complete the application design requirements for usability testing. Finally, the user would agree that the system satisfies the criteria in user acceptance testing, is usable, and can be deployed. In some lifecycle processes, usability may be the last phase before release. For example, some organizations may have a beta test to ensure that the conditions mimic real-world applications before the actual test. They would then evaluate the results to determine if changes need to be made before real testing.

Different tests may include the A/B test, prototype test, formative usability test, summative usability test, eye-tracking test, and questionnaires. Each test evaluates and assesses whether users are happy with the application, and this application will add value by input from the user.

The A/B test is used to test web pages. There would be two designs designated A and B, and users would test both designs. The information would be gathered and evaluated. The A/B test would be particularly useful for user-centric testing. The test involves two options, A and B. Users are selected and put into Group A or B. Then, each group is given a different scenario to test based on a different version of the same application. For example, Group A may interface with all menu items in a vertical row, whereas Group B has the same information in a horizontal row. After the test, both results are compared based on what each group said about the layout of the interface. This approach provides the designers and developers with a unique approach to which design would be the most functional for the application. Based on the results, there may be a collaboration of both results to finalize one design. See Figure 3.

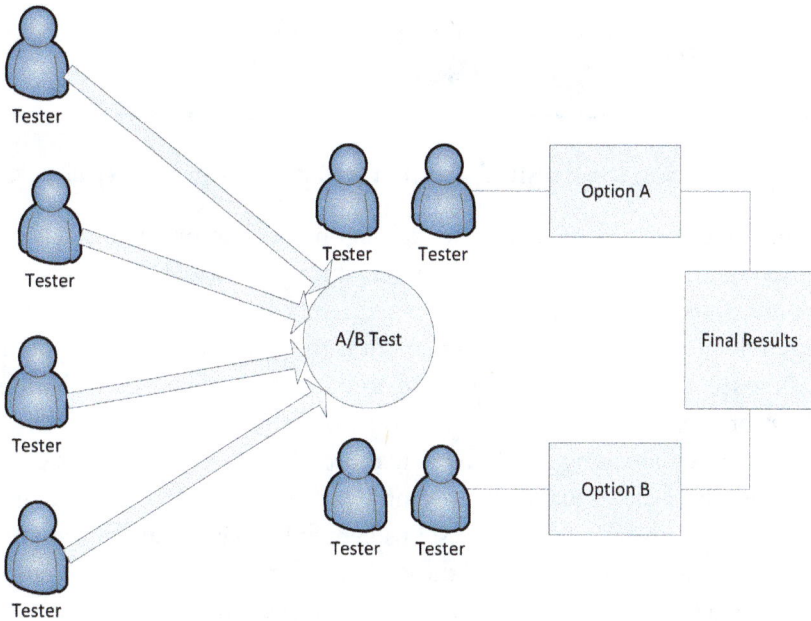

Figure 3. A/B Testing

Prototype testing involves the user and the designer, who may utilize a tool to show the different steps in the development by designing the application based on the requirements. Then, the prototype development is used to see whether this meets their goals and whether they need to reevaluate the design.

The formative usability test focuses more on the quality and whether the application meets acceptance testing goals. Therefore, formative usability testing may be considered the baseline for testing. Then there is a test called Summative usability. Summative usability testing is a test process that happens later in the lifecycle development process, and users may work together to see whether the design meets their objectives.

Then there is Eyetracking, which uses the camera to play a role, and users' eye movements are tracked. The developer and designer may find this helpful when examining the time it takes to complete a task and whether too much movement would negatively impact the user. Finally, questionnaires are used by designers, developers, and users to improve development, manage development, and even use for best practices (Stumpf, 2018).

Critical Thinking:

Why should developers not be the only testers?

Implementing the finished product and providing quality checks

Since our society is increasingly moving into online organizational applications, virtual teaching, and online education, virtual classes are becoming more and more effective in instructing students. Because of this, there is more and more movement from the face-to-face development of classes, applications, and products to the virtual realm. Mastrolembo, Castronovo, and Ciribini (2020) discussed how VR (Virtual Reality) systems are increasing and becoming more of the norm than the exception. So there are virtual reality prototypes developed, placed online, and tested and implemented. The systems benefit by being visually aesthetic and also interactive. Because of the design and system requirements, prototypes become increasingly important to develop before implementing the systems in a real-world setting.

An organization should develop a plan, design, and development process that works in a virtual and face-to-face environment. Many virtual environments can help designers and developers work from development to implementation.

Vandenberg, Harmann, and deGraaf (2017) noted that when evaluating the prototype, that "enables clients to navigate through and comment on a design-in-progress individually" (p. 305). Therefore, evaluation is necessary before implementing the system or application.

Implementing an application will require the user and developer, and they need to discuss the roles that each will have and the tasks that will be carried out. Following are several suggested steps:

1. First, review the actual plan and how the design should look. Ask the user to ensure this was the agreed plan before beginning the project.

2. Evaluate the application that will be implemented and compare it to the prototype design. Then, validate with the user that the application compares with the prototype. Ask if they see any differences.

3. Discuss with all stakeholders/users whether the business requirements have been met and consider the changes requested after testing the prototype.

4. Evaluate the specifications and ensure that the servers handle the design and development documents load.

5. Implement and move the application to production. Review how the application runs compared to the prototype. Develop a timeline that will work with the users to sync the data with the old or existing system. Finally, ask the user to test the system and work with experts to validate testing approaches to validate reliability.

6. Evaluate the results and ensure the information is valid and verified before going live with the application. Ask the user to validate the information that is received from the application.

7. Continue to monitor the system after going live and apply a continuous improvement process. Have the user involved in continual evaluation and improvements.

8. Make sure that everyone understands the features of the new application and how to report errors if they are found in a support group. Provide users with the support process.

Tools can be helpful when developing a process to test, so it is essential to look at the many different techniques and tools. Requirements testing ensures that the user's requirements are met and contribute to the overall operating performance. Also, testing improves the overall quality of the software.

There is no specific design applicable to software development lifecycles like Agile and Spiral when setting up a requirements test. Instead, the user approved a list of requirements and implemented by the developer. The list can be developed into test cases and scripts (dos, S. J., Martins, L. E., G., de Santiago Júnior Valdivino, A., Povoa, L. V., & dos Santos Luciana Brasil, R., 2020). Separating the functional from the non-functional requirements provides an understanding of whether it is a functional issue with the software or an error with the system. For example, security would be more interested in the results from non-functional to determine whether there are security issues that need to be addressed than the actual function, which is the user's concern.

At the same time, the users consider the time it takes to navigate to a particular part of the program, which would deal with the functional aspect of the application. Requirements tools are available that are open-source software helpful in gathering requirements. These tools are used to help coordinate the requirements. The following are tools for requirements management:

- Arbiter - http://arbiter.sourceforge.net/
- Archi - https://www.archimatetool.com/
- Ephemeris - https://github.com/shuart/ephemeris
- JavaRequirements Tracer - http://reqtracer.sourceforge.net/
- Requirement Heap - https://sourceforge.net/projects/reqheap/

Chapter 8 will focus on the maintenance and continuous improvement process.

Reflections – Chapter 7

Automated versus manual

Establishing a test process and procedure will help evaluate the application. Reporting the errors and establishing a log will improve the process of fixing any mistakes. If working through the application manually, specific areas will focus on the user. When using an automated tool, the automation may be set to evaluate specific areas, depending on what the user and developer are evaluating.

Questions to consider:

1. When determining a test approach, how does a user select the process?

2. If a user finds errors, should they determine whether it is necessary to improve or if the application works ok without improvements?

MAINTENANCE AND CONTINUOUS IMPROVEMENT

Maintenance and continuous improvement

Maintenance is a significant part of software development and ongoing continuous improvement. Sujay and Reddy (2017) suggested different types of maintenance for users and developers. The different approaches to maintenance can be:

- Corrective Maintenance

- Adaptive Maintenance

- Perfective Maintenance

- Preventive Maintenance

Each type of maintenance requires a different approach when managing applications and improving the operational improvement of the application. For example, corrective action ensures that changes and updates are applied to fix issues that may make the software unusable when evaluating an application.

Adaptive maintenance incorporates changes helping users keep up with new and innovative changes. Perfective Maintenance is based on updates that may need to be adjusted and keep the application working correctly and incorporate usable software. Finally, Preventative Maintenance is useful and used in most organizations. and works towards making sure the software is correct before issues occur (Sujay & Reddy, 2017).

Collaboration with team members

Laubheimer (2016) noted that UX professionals manage and collaborate on deliverables to be successful. As mentioned, collaboration efforts are made using different tools and software, such as prototype applications like wireframes. When working in groups, teams provide a better application than those working alone (Laubheimer, 2016). Also, when team members work with users, many other opinions and ideas are generated, improving the overall performance by including the necessary elements required.

Since moving towards the internet, collaborating is critical, and picking a collaborative tool will improve the process and create a community where specific topics are discussed and managed using a continual improvement process. Following are several tools that will be beneficial:

- Robin is a tool used to set up a hybrid workplace (robinpowered. com)

- MITeam Meetings is a real-time video conferencing tool (mitel.com)

- Bluescape is a tool for running a collaborative meeting (create.bluescape.com)

- Webex is a hybrid video conferencing tool (webex.com)

- Avaya is a cloud communication tool to work with teams (avaya.com)

Working in groups, particularly with remote interaction, provides unique challenges. One of the biggest challenges is how collaboration will be set up and when the collaboration will occur. Continuous improvement can be timely but should be carried out by a user or developer.

> **Critical Thinking:**
> **What does the future hold for user centric design?**

Continuous Improvement

Continuous improvement occurs when managing incremental changes that help an application's functionality while meeting users' needs. A helpful technique when going through continuous improvement is the KAIZEN process. The KAIZEN process is considered a continuous improvement process developed to provide successive steps for working through improvements.

Singh and Singh (2012) noted that KAIZEN has two concepts: KAI, which means change, and ZEN, which means for the betterment of the application. KAIZEN has been interpreted in many different ways, and one way is using the following steps, in which each step builds upon the previous step.

It is essential first to understand and manage the tasks. Then put those tasks into an order that makes sense. This will help achieve the best approach and implement the solution. Then there needs to be a review and an analysis to maximize solutions.

It is clear by looking at the KAIZEN steps for continual improvement; the user plays a crucial role in helping to make continuous improvements. For example, when an application is deployed, the user will work with the application to accomplish tasks. As they work through the use of the application, they may see areas where possible changes could be made to improve the efficiency and overall success of the application's operation.

The KAIZEN process can be modified and changed to include critical steps that would be beneficial and help users maintain a sound and operational system based on the organization's overall plan and objectives. For example, the management of sort, set in order, shine, standardize, and sustain provides a basic step-by-step process to help maximize efficiency and minimize errors in operation.

Figure 1 indicates that user-centric design and development encompasses usability, accessibility, reliability, desirability, speed, stability, and security. Each part elicits information from the user and involves the user in the development process. For User-Centric design to be successful, understanding the behavior and experience of the user is based on behavior and ethical implications that may impact the design.

Strategies that assess users' behavior and feedback can be valuable tools for improving users' products and services. The user's perspective is critical since they are the ones that will ultimately use the application, and their collaboration with the developers will be beneficial in understanding the requirements and design. In addition, the users significantly impact the organization, and as part of the team, dynamics can jettison the application to its success.

Future direction

So what does the future of user-centricity hold for developing applications, software, and products? First, organizations must consider the factors that revolve around the user since this is the focus of application and product success. The user needs to help define and participate in the overall development process. Figure 1 is a proposed model that includes the user in the overall development process and suggests that the user's skills and input are further applied.

In Figure 1, the user-centric design focuses on design, development, usability, accessibility, reliability, desirability, speed, stability, and security. So it takes more than one action to create a functional design that users feel is based on meeting the goals and objectives. The user must understand that their behavior and experience are both areas that are beneficial to the developer. Also, there are ethical implications of psychology built into the design. A designer and developer need to understand the psychological approach to design before determining the strategies to assess the behavior and make the design useable. When designers and developers receive user feedback for the application, they can incorporate the user's perspective into the design and create a collaborative approach to implementing the application. An excellent way to keep track of suggestions and corrections would be to add a tracking tool that provides specific dates and times to manage the corrections. Users play a crucial role and impact the organization through collaboration and cooperation for the successful application of the design.

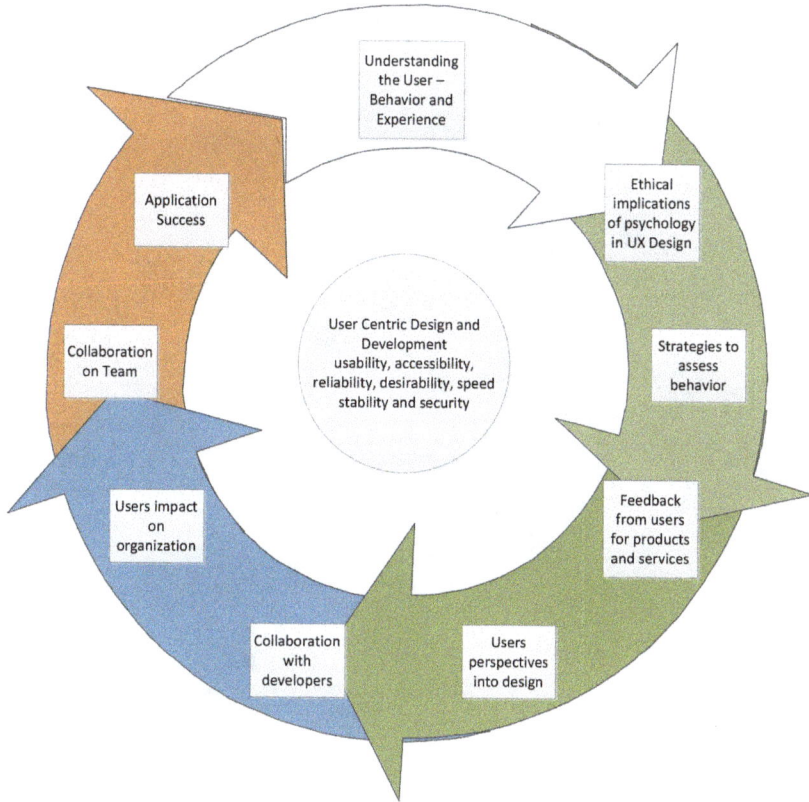

Figure 1. User-Centric Design

Reflection Chapter 8

Who is responsible for continual improvement and maintenance?

When setting up a maintenance process, does the organization need to determine whether Corrective Maintenance, Adaptive Maintenance, Perfective Maintenance, or Preventive Maintenance is used? Again, the user and the developer should provide input to determine how they move toward maintenance.

Questions:

1. How does Corrective Maintenance improve the application?

2. Would Preventative Maintenance help improve the user's interaction?

REFERENCES

Al-Megren, Khabti, J., & Al-Khalifa, H. S. (2018). A Systematic Review of Modifications and Validation Methods for the Extension of the Keystroke-Level Model. *Advances in Human-Computer Interaction, 2018,* 1–26. https://doi.org/10.1155/2018/7528278

Aufderheide, Patricia. Ethical Challenges for Documentarians in a User-Centric Environment (October 3, 2013). Forthcoming in: New Documentary Ecologies, Palgrave, Available at SSRN: https://ssrn.com/abstract=2335497

Babich. (2019). What is rapid prototyping? Retrieved from https://xd.adobe.com/ideas/process/prototyping/rapid-prototyping-efficient-way-communicate-ideas/

Balsamiq. (2021). How to start a wireframe project. Retrieved from https://balsamiq.com/learn/articles/how-to-start-a-wireframe/

Baskar, S. (2017). 6 reasons designing UX for IoT is so difficult. Machine Design, Retrieved from https://www.proquest.com/trade-journals/6-reasons-why-designing-ux-iot-is-so-difficult/docview/1925074786/se-2?accountid=8289

Berkeley Web Access. (2021). Types of Assistive Technology, Retrieved from https://webaccess.berkeley.edu/resources/assistive-technology

Bhaskar, P. R., Jukan, A., Katsaros, D., & Goeleven, Y. (2011). Architectural requirements for cloud computing systems: An enterprise cloud approach. *Journal of Grid Computing, 9*(1), 3-26. doi:http://dx.doi.org/10.1007/s10723-010-9171-y

Bowler, L., Koshman, S., Oh, J. S., He, D., Callery, B. G., Bowker, G., & Cox, R. J. (2011). Issues in user-centered design in LIS. *Library Trends, 59*(4), 721-752. doi:http://dx.doi.org/10.1353/lib.2011.0013

Byerley, S. L., & Chambers, M. B. (2002). Accessibility and usability of web-based library databases for non-visual users. *Library Hi Tech, 20*(2), 169. Retrieved from https://www-proquest-com.ezproxy1.apus.edu/scholarly-journals/accessiblity-usability-web-based-library/docview/200616417/se-2?accountid=8289

Coyle. (2021). 3 techniques to influence user behavior. Retrieved from https://uxmag.com/articles/3-techniques-to-influence-user-behavior

Dar, H., Lali, M. I., Ashraf, H., Ramzan, M., Amjad, T., & Shahzad, B. (2018). A Systematic Study on Software Requirements Elicitation Techniques and its Challenges in Mobile Application Development. *IEEE Access*, 6, 63859–63867. https://doi.org/10.1109/ACCESS.2018.2874981

Derby, J. (2019, November 18). The good & bad of website design. University Wire Retrieved from https://www.proquest.com/wire-feeds/good-amp-bad-website-design/docview/2315303495/se-2?accountid=8289

dos, S. J., Martins, L. E., G., de Santiago Júnior Valdivino, A., Povoa, L. V., & dos Santos Luciana Brasil, R. (2020). Software requirements testing approaches: A systematic literature review. *Requirements Engineering*, 25(3), 317-337. doi:http://dx.doi.org/10.1007/s00766-019-00325-w

Experience, Abu (2018). Psychology Principles in UX Design. Retrieved from https://uxbert.com/ux-psychology-principles-design-ux/

Følstad, A. (2017). Users' design feedback in usability evaluation: A literature review. *Human-Centric Computing and Information Sciences*, 7(1), 1-19. doi:http://dx.doi.org.ezproxy1.apus.edu/10.1186/s13673-017-0100-y

Fritton, Cheverst, Kray, Dix, Funcefield & Saslis-Lagoudakis. (2005). Rapid prototyping and user-centered design of interactive display-based systems. Pervasive Computing. Retrieved from www.computer.org/pervasive.

Guru99. (2021). V-Model in Software Testing. Retrieved from https://www.guru99.com/v-model-software-testing.html

Henderson, K. (1991). "Flexible Sketches and Inflexible Data Bases: Visual Communication, Conscription Devices, and Boundary Objects in Design Engineering." *Sci., Technol. Human Values*, 16(4), pp. 448–473.

Hibbeln, Martin & Jenkins, Jeffrey & Schneider, Christoph & Valacich, Joseph & Weinmann, Markus. (2017). How Is Your User Feeling?

Inferring Emotion Through Human-Computer Interaction Devices. *MIS Quarterly*, 41. DOI: 10.25300/MISQ/2017/41.1.01.

Interaction Design Foundation. (n.d). Fitts' Law. Retrieved from https://www.interaction-design.org/literature/topics/fitts-law

Interaction Design Foundation. (n.d). Fitts's Law: The Importance of Size and Distance in UI Design. Retrieved from https://www.interaction-design.org/literature/article/fitts-s-law-the-importance-of-size-and-distance-in-ui-design

Interaction Design Foundation. (n.d). Design Thinking. Retrieved from https://www.interaction-design.org/literature/topics/design-thinking

Jia, J., & Capretz, L. F. (2018). Direct and mediating influences of user-developer perception gaps in requirements understanding on user participation. *Requirements Engineering*, 23(2), 277-290. doi:http://dx.doi.org.ezproxy2.apus.edu/10.1007/s00766-017-0266-x

Jongsoo Ha, & Sungwoo An. (2017). Analysis of Eye Movements for Designing a Portal Service based on the User Experience. International Information Institute (Tokyo). *Information*, 20(8A), 5655–5660.

Kaasinen, E., Kymäläinen, T., Niemelä, M., Olsson, T., Kanerva, M., & Ikonen, V. (2012). A User-Centric View of Intelligent Environments: User Expectations, User Experience and User Role in Building Intelligent Environments. *Computers* (Basel), 2(1), 1–33. https://doi.org/10.3390/computers2010001

Laubheimer (2016). How Ux professionals collaborate on deliverables. Retrieved from https://www.nngroup.com/articles/ux-deliverables-collaboration/

Lauff, C.A., Kotys-Schwartz, D. & Rentschler, M.E., (2018). What is a prototype: What are the roles of prototypes in companies? *Journal of Mechanical Design*, DOI: 10.1115/1.4039340

Mastrolembo, Ventura S., Castronovo, F., Ciribini, A.L.C. (2020). A design review session protocol for the implementation of immersive virtual reality in usability-focused analysis, *ITcon* Vol. 25, Special issue eWork and eBusiness in Architecture, Engineering and Construction

2018, pg. 233-253, https://doi.org/10.36680/j.itcon.2020.014

Model-Based Design - Cognitive (User) Models Retrieved from http://www.cs.umd.edu/class/fall200

Myers, K. L., Tyson, W. M., Wolverton, M. J., Jarvis, P. A., Lee, T. J., & desJardins, M. (2002, October). PASSAT: A user-centric planning framework. Proceedings of the 3rd International NASA Workshop on Planning and Scheduling for Space (pp. 1-10).

Pibernik, J., Dolic, J., Hrvoje, A. M., & Kanizaj, B. (2019). The effects of the floating action button on quality of experience. *Future Internet,* 11(7), 148. doi:http://dx.doi.org/10.3390/fi11070148

QRA. (2021). Functional vs. Non-functional requirements: The Definitive Guide. Retrieved from https://qracorp.com/functional-vs-non-functional-requirements/

Rakheja, J. (2018, October 29). Motion design is the SFX of UX today. PCQuest, Retrieved from https://www.proquest.com/magazines/motion-design-is-sfx-ux-today/docview/2138240668/se-2?accountid=8289

Ren. (2020). Design of Mobile APP User Behavior Analysis Engine Based on Cloud Computing. *Journal of Physics.* Conference Series, 1533(2), 22092–. https://doi.org/10.1088/1742-6596/1533/2/022092

ReQtest. (2021). Why is the difference between functional and Non-functional requirements important? Retrieved from https://reqtest.com/requirements-blog/functional-vs-non-functional-requirements/

Rochford, J. (2019). Accessibility and IoT / smart and connected communities. *AIS Transactions on Human-Computer Interactions,* 11(4), 253-263. doi:http://dx.doi.org/10.17705/1thci.00124

Sarker, I. H., Colman, A., Han, J., Khan, A. I., Abushark, Y. B., & Khaled, S. (2020). BehavDT: A behavioral decision tree learning to build user-centric context-aware predictive model. *Mobile Networks and Applications,* 25(3), 1151-1161. doi:http://dx.doi.org/10.1007/s11036-019-01443-z

Singh, J., & Singh, H. (2012). Continuous improvement approach: State-of-art review and future implications. *International Journal of*

Lean Six Sigma, 3(2), 88-111. doi:http://dx.doi.org/10.1108/204014
61211243694

Sohaib, O., Hussain, W., & Badini, M. K. (2011). User experience (UX) and web accessibility standards. *International Journal of Computer Science Issues* (IJCSI), 8(3), 584-587. Retrieved from https://www. proquest.com/scholarly-journals/user-experience-ux-web-accessibil ity-standards/docview/873265725/se-2?accountid=8289

Stumpf, C. (2018). 6 usability testing methods that will improve your software. InfoWorld.Com, Retrieved from https://www.proquest. com/trade-journals/6-usability-testing-methods-that-will-improve/ docview/2071962296/se-2?accountid=8289

Su, Li, F., Shi, G., Geng, K., & Xiong, J. (2016). A User-Centric Data Se-cure Creation Scheme in Cloud Computing. *Chinese Journal of Elec-tronics*, 25(4), 753–760. https://doi.org/10.1049/cje.2016.07.017

Sujay, V., & M.B.R. (2017). Advanced architecture-centric software maintenance. *I-manager's Journal on Software Engineering*, 12(1), 1-5. doi:http://dx.doi.org/10.26634/jse.12.1.13917

TechTarget. (2021). Spiral Model. Retrieved from https://searchsoft-warequality.techtarget.com/definition/spiral-model

Urquhart, L. (2018). Ethical dimensions of user-centric regulation. *SIG-CAS Comput. Soc.*, 47, 81-95.

Usability.gov. (2021). User-centered design basics. Retrieved from https://www.usability.gov/what-and-why/user-centered-design.html

Usability.gov. (2021). User-Centered Design Process Map. Retrieved from https://www.usability.gov/how-to-and-tools/resources/ucd-map.html

Uxdesignworld. (2020). Good design vs. bad design. Retrieved from https://uxdworld.com/2020/02/22/good-design-vs-bad-design/

Marc van den Berg, Timo Hartmann, & Robin de Graaf. (2017). Sup-porting design reviews with pre-meeting virtual reality environments. *Journal of Information Technology in Construction* (ITcon), Vol. 22, pg. 305-321, http://www.itcon.org/2017/16

Venter, C. (2019). A critical systems approach to elicit user-centric business intelligence business requirements. *Systemic Practice and Action Research*, 32(5), 481-500. doi:http://dx.doi.org.ezproxy2.apus.edu/10.1007/s11213-018-9468-5

W3C Web Accessibility Initiative. (2021). Retrieved from https://www.w3.org/WAI/redesign/ucd

Wentz, Dung, & Tressler. (2017). Exploring the accessibility of banking and finance systems for blind users. Retrieved from https://firstmonday.org/ojs/index.php/fm/article/view/7036/5922

Whatis.com. (2021). Hick's Law. Retrieved from https://whatis.techtarget.com/definition/Hicks-law

Related Titles from Westphalia Press

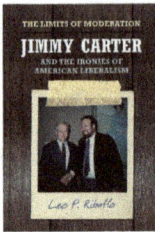

The Limits of Moderation: Jimmy Carter and the Ironies of American Liberalism
by Leo P. Ribuffo

The Limits of Moderation: Jimmy Carter and the Ironies of American Liberalism is not a finished product. Yet, this book is a close and careful history of a short yet transformative period in American political history, when big changes were afoot. and continue to shape our world.

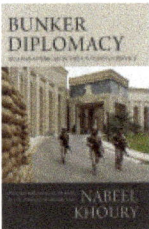

Bunker Diplomacy: An Arab-American in the U.S. Foreign Service
by Nabeel Khoury

After twenty-five years in the Foreign Service, Dr. Nabeel A. Khoury retired from the U.S. Department of State in 2013 with the rank of Minister Counselor. In his last overseas posting, Khoury served as deputy chief of mission at the U.S. embassy in Yemen (2004-2007).

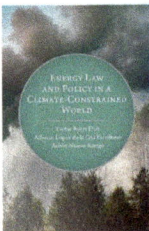

Energy Law and Policy in a Climate-Constrained World
by Victor Byers Flatt, Alfonso López de la Osa Escribano, Aubin Nzaou-Kongo

This book presents reflections on concepts, foreign policy, regional and international cooperation, and the specific role the state is to play when it comes to such thing as energy law and policy.

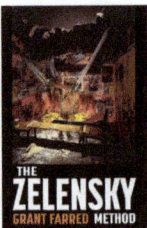

The Zelensky Method
by Grant Farred

Locating Russian's war within a global context, The Zelensky Method is unsparing in its critique of those nations, who have refused to condemn Russia's invasion and are doing everything they can to prevent economic sanctions from being imposed on the Kremlin.

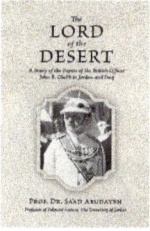

The Lord of the Desert: A Study of the Papers of the British Officer John B. Glubb in Jordan and Iraq
by Dr. Sa'ad Abudayeh

John Bajot Glubb, a British engineer officer, was sent to Iraq in 1920 to resolve the problems which erupted after the Iraqi revolt. He remained in the area for ten years, working with the Bedouins. In 1930, he moved to Jordan for twenty-six successful years. He invented what Dr. Abudayeh calls the Diplomacy of Desert.

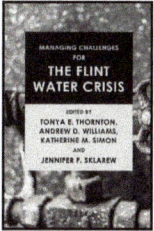

Managing Challenges for the Flint Water Crisis
Edited by Toyna E. Thornton, Andrew D. Williams, Katherine M. Simon, Jennifer F. Sklarew

This edited volume examines several public management and intergovernmental failures, with particular attention on social, political, and financial impacts. Understanding disaster meaning, even causality, is essential to the problem-solving process.

Resistance: Reflections on Survival, Hope and Love
Poetry by William Morris, Photography by Jackie Malden

Resistance is a book of poems with photographs or a book of photographs with poems depending on your perspective. The book is comprised of three sections titled respectively: On Survival, On Hope, and On Love.

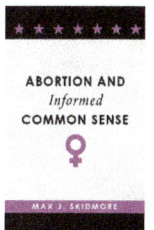

Abortion and Informed Common Sense
by Max J. Skidmore

The controversy over a woman's "right to choose," as opposed to the numerous "rights" that abortion opponents decide should be assumed to exist for "unborn children," has always struck me as incomplete. Two missing elements of the argument seems obvious, yet they remain almost completely overlooked.

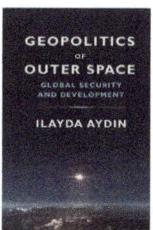

Geopolitics of Outer Space: Global Security and Development
by Ilayda Aydin

A desire for increased security and rapid development is driving nation-states to engage in an intensifying competition for the unique assets of space. This book analyses the Chinese-American space discourse from the lenses of international relations theory, history and political psychology to explore these questions.

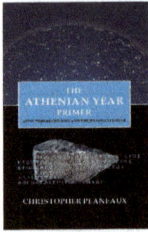

The Athenian Year Primer: Attic Time-Reckoning and the Julian Calendar
by Christopher Planeaux

The ability to translate ancient Athenian calendar references into precise Julian-Gregorian dates will not only assist Ancient Historians and Classicists to date numerous historical events with much greater accuracy but also aid epigraphists in the restorations of numerous Attic inscriptions.

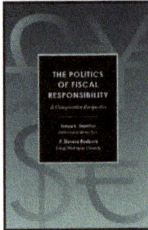

The Politics of Fiscal Responsibility: A Comparative Perspective
by Tonya E. Thornton and F. Stevens Redburn

Fiscal policy challenges following the Great Recession forced members of the Organisation for Economic Co-operation and Development (OECD) to implement a set of economic policies to manage public debt.

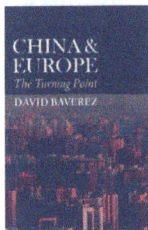

China & Europe: The Turning Point
by David Baverez

In creating five fictitious conversations between Xi Jinping and five European experts, David Baverez, who lives and works in Hong Kong, offers up a totally new vision of the relationship between China and Europe.

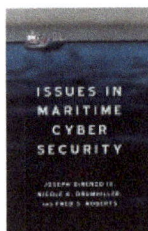

Issues in Maritime Cyber Security
Edited by Dr. Joe DiRenzo III, Dr. Nicole K. Drumhiller, and Dr. Fred S. Roberts

The complexity of making MTS safe from cyber attack is daunting and the need for all stakeholders in both government (at all levels) and private industry to be involved in cyber security is more significant than ever as the use of the MTS continues to grow.

Freemasonry, Heir to the Enlightenment
by Cécile Révauger

Modern Freemasonry may have mythical roots in Solomon's time but is really the heir to the Enlightenment. Ever since the early eighteenth century freemasons have endeavored to convey the values of the Enlightenment in the cultural, political and religious fields, in Europe, the American colonies and the emerging United States.

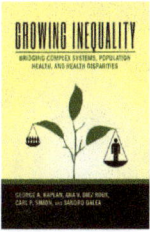

Growing Inequality: Bridging Complex Systems, Population Health, and Health Disparities
Editors: George A. Kaplan, Ana V. Diez Roux, Carl P. Simon, and Sandro Galea

Why is America's health is poorer than the health of other wealthy countries and why health inequities persist despite our efforts? In this book, researchers report on groundbreaking insights to simulate how these determinants come together to produce levels of population health and disparities and test new solutions.

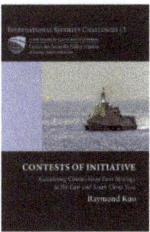

Contests of Initiative: Countering China's Gray Zone Strategy in the East and South China Seas
by Dr. Raymond Kuo

China is engaged in a widespread assertion of sovereignty in the South and East China Seas. It employs a "gray zone" strategy: using coercive but sub-conventional military power to drive off challengers and prevent escalation, while simultaneously seizing territory and asserting maritime control.

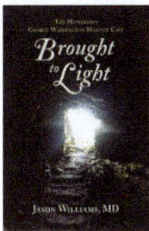

Brought to Light: The Mysterious George Washington Masonic Cave
by Jason Williams, MD

The George Washington Masonic Cave near Charles Town, West Virginia, contains a signature carving of George Washington dated 1748. Although this inscription appears authentic, it has yet to be verified by historical accounts or scientific inquiry.

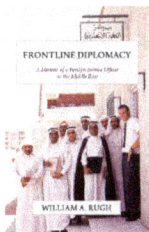

Frontline Diplomacy: A Memoir of a Foreign Service Officer in the Middle East
by William A. Rugh

In short vignettes, this book describes how American diplomats working in the Middle East dealt with a variety of challenges over the last decades of the 20th century. Each of the vignettes concludes with an insight about diplomatic practice derived from the experience.

westphaliapress.org

Policy Studies Organization

The Policy Studies Organization (PSO) is a publisher of academic journals and book series, sponsor of conferences, and producer of programs.

Policy Studies Organization publishes dozens of journals on a range of topics, such as European Policy Analysis, Journal of Elder Studies, Indian Politics & Polity, Journal of Critical Infrastructure Policy, and Popular Culture Review.

Additionally, Policy Studies Organization hosts numerous conferences. These conferences include the Middle East Dialogue, Space Education and Strategic Applications Conference, International Criminology Conference, Dupont Summit on Science, Technology and Environmental Policy, World Conference on Fraternalism, Freemasonry and History, and the Internet Policy & Politics Conference.

For more information on these projects, access videos of past events, and upcoming events, please visit us at:

www.ipsonet.org

www.ingramcontent.com/pod-product-compliance
Lightning Source LLC
Chambersburg PA
CBHW060023210326
41519CB00042BA/6916